Seed Priming Approaches That Achieve Environmental Stress Tolerance

Seed Priming Approaches That Achieve Environmental Stress Tolerance

Guest Editors

José Antonio Hernández Cortés
Gregorio Barba-Espín
Pedro Diaz-Vivancos

MDPI

Basel • Beijing • Wuhan • Barcelona • Belgrade • Novi Sad • Cluj • Manchester

Guest Editors
José Antonio Hernández Cortés
Group of Fruit Biotechnology, Department of Fruit Breeding, CEBAS-CSIC, Murcia
Spain

Gregorio Barba-Espín
Centro de Edafología y Biología Aplicada del Segura, CSIC, Grupo de Biotecnología de Frutales, Departamento de Mejora Vegetal
Murcia
Spain

Pedro Diaz-Vivancos
CEBAS-CSIC, Department of Plant Breeding, Campus de Espinardo
Murcia
Spain

Editorial Office
MDPI AG
Grosspeteranlage 5
4052 Basel, Switzerland

This is a reprint of the Special Issue, published open access by the journal *Seeds* (ISSN 2674-1024), freely accessible at: https://www.mdpi.com/journal/seeds/special_issues/Seed_Priming.

For citation purposes, cite each article independently as indicated on the article page online and using the guide below:

Lastname, A.A.; Lastname, B.B. Article Title. *Journal Name* **Year**, *Volume Number*, Page Range.

ISBN 978-3-7258-2835-7 (Hbk)
ISBN 978-3-7258-2836-4 (PDF)
https://doi.org/10.3390/books978-3-7258-2836-4

Contents

About the Editors

José Antonio Hernández Cortés

José Antonio Hernández Cortés is a Senior Researcher from the Spanish National Research Council (CSIC) in the Research Institute "Centre for Soil Science and Applied Biology of Segura" (CEBAS) in Murcia (Spain) and an external teacher of Plant Physiology at the Catholic University San Antonio in Murcia (UCAM, Spain). Much of his scientific career has been dedicated to studying the response of plants to salinity. During his Ph.D. period, he studied the effects of salt stress on antioxidant and ROS (reactive oxygen species) metabolism at the subcellular level. After his postdoctoral stays in Toulouse (ENSAT, France) and Norwich (John Innes Centre, UK), he continued investigating the involvement of ROS in the effect of salinity on the activity and expression of ASC-GSH cycle enzymes and SOD isoenzymes. Then, he started a new research line related to the physiological and biochemical bases of the response to biotic stress in *Prunus* sp. In parallel, he studied the role of redox metabolism on seed germination, reporting that H2O2 promoted germination and early seedling growth in a concentration-dependent manner and observing an interaction among the redox state and hormone profile, orchestrated by H2O2, in the induction of proteins related to plant signaling and development during the early growth of pea seedlings. Recently, he has started a new line of research about the effect of halophyte-based management in the physiological and biochemical responses of crop plants under saline conditions (PRIMA Programme, EU). In addition, he has transfered the results of his research to the private sector through research projects with private companies, focused on the agronomic management of citrus and fruit trees. Hernández is Deputy Director of his research institute (CEBAS-CSIC) and Head of the Plant Breeding Department. He has published about 130 papers in SCI journals, mostly Q1, with an h index of 50 (WOS), and has more than 11,000 citations.

Gregorio Barba-Espí

Gregorio Barba-Espí is a plant biologist widely dedicated to understanding and improving plant development, abiotic stress response, and seed germination through physiology, biochemistry, and biotechnology studies. Since the beginning of his research career, his scientific outputs have aimed to advance scientific knowledge, as well as to be applied within society, especially within the agri-food sector. He holds a degree in Biology (University of Murcia, 2006) and conducted his Ph.D. studies at CEBAS-CSIC (2008-2011), focusing on characterization of the antioxidant metabolism of pea plants during germination and plant growth. His international trajectory comprises two pre-doctoral stays at Bayer CropScience (Lyon) and at the University of Leeds, and 6 years in Denmark as a postdoctoral researcher, first at the Technical University of Denmark (20 months) and later at the University of Copenhagen (46 months, of which 20 months were as an Assistant Professor and PI of a project). He holds over 50 SCI publications and has an h index of 25. He is the main inventor of two European patent applications (2018 and 2019). He has conducted research in seven different departments and co-authored papers with over 100 scientists. The internalization of his research is reflected in 30 communications to international conferences (six oral communications). He has solid editorial activity (Guest Editor of four Special Issues in Q1 journals, co-editor of a book, and a reviewer for numerous Q1 journals). He has organized R&D activities, such as an international workshop. Since 2017, he has supervised 19 students including three doctoral and five MSc candidates, who are now active investigators in the public or private sectors. His current research (2020-present) involves the establishment of halophyte-based saline agriculture, which includes in vitro culture as a platform for the propagation of elite halophyte germplasm.

Pedro Diaz-Vivancos

Pedro Diaz-Vivancos, During his PhD, Pedro Diaz-Vivancos carried out several studies on the response of Prunus and pea plants against *Plum Pox virus* (PPV) infection at the subcellular level. His research team demonstrated for the first time that PPV produced oxidative stress in infected plants by inducing an imbalance in antioxidant systems as well as resulting in increased generation of reactive oxygen species (ROS) in chloroplasts, suggesting that chloroplasts can be a source of oxidative stress during viral disease development. In 2008-2009, he received a postdoctoral grant from Christine H. Foyer´s lab, during which his research activity focused on the role of glutathione (GSH) in the regulation of cell proliferation and plant development. His team demonstrated that, similarly to in animal cells, in the early stages of plant cell proliferation, GSH is recruited into the nucleus. This has a profound effect on the whole-cell redox state and on gene expression, mainly affecting the cellular oxidative defense shield. He was then awarded with a postdoctoral position at CEBAS-CSIC. During this period, he worked on three main topics: (1) the effect of H2O2 on seed germination and seedling establishment, (2) induced tolerance against abiotic stress in plum plants by increasing their antioxidant capacity, and (3) the description of a new pathway for salicylic acid (SA) biosynthesis in peach, linking cyanogenic glycosides turnover and the SA biosynthetic pathway. In 2018, he joined the University of Murcia as a lecturer in the Plant Biology Department. In 2020, he became a Senior Researcher at CEBAS-CSIC. Within the frame of a PRIMA project, he has studied the effect of halophyte-based crop management in tomato at the physiological, biochemical, and metabolomic levels. In addition, his team are conducting research studies with agricultural companies with the aim of reducing both fertilizer and water inputs without affecting crop production. As a result of his research work, he has published more than 70 manuscripts in SCI journals, mostly Q1, accounting for more than 4500 citations.

Preface

Seed priming offers numerous benefits that contribute to sustainable and efficient crop production. This technique enhances seed performance and leads to faster and more uniform germination, which is crucial for establishing healthy crops. One of the primary advantages of seed priming is its ability to improve germination rates, especially under environmental stress conditions such as drought, salinity, hypoxia, or low and high temperatures. By supporting the germination process, primed seeds can establish themselves quickly and robustly, which reduces the vulnerability of seedlings to these stress factors. Additionally, seed priming can lead to improved crop yields. With more uniform germination, plants grow more consistently, leading to better competition with weeds and more efficient use of water and nutrients. This uniformity in seed germination can also facilitate a more effective timing of agricultural practices, such as irrigation and fertilization, ultimately enhancing overall productivity. Seed priming can be an environmentally friendly practice, as it often reduces the need for chemical seed treatments. By enhancing the natural resilience and vigor of seeds, farmers can potentially reduce their reliance on synthetic pesticides and fertilizers, leading to more sustainable farming practices.

Seed priming techniques are closely linked to both seed technology and seed biology. Understanding the interplay between these components is essential for optimizing agricultural practices. Seed technology encompasses the tools and techniques used to enhance seed quality and performance. It plays a pivotal role in the seed priming process by providing the methods and materials necessary for effective treatment. These techniques include strategies such as hydropriming, osmopriming, chemopriming, biopriming, and hormonal priming that are adapted to different types of crops and environmental conditions, ensuring that seeds have the best possible start in the germination process. Seed biology involves understanding the physiological, biochemical, and molecular processes that occur during seed development, dormancy, germination, and vigor. Knowledge of seed biology is fundamental in optimizing seed priming techniques. In addition it is important to underline that priming could reduce the longevity of seeds; it is therefore important to maintain the benefit of the technique by storing seeds in good conditions or timing the priming of the seeds so it is not carried out too long before sowing.

In the current era, with a continuously increasing population, the consequences of climate change may compromise food security and increase the rate of hunger in the world. In that regard, seed priming offers a promising strategy to mitigate the effects of climate change on crop production. By enhancing seed germination, seedling growth, and stress tolerance, seed priming can contribute to more resilient agricultural systems and help ensure food security for future generations. Continued research and innovation in this area are essential to maximize the potential benefits of seed priming and adapt to the changing climate.

Thus, the success of seed priming depends on the seamless integration of seed technology and seed biology. By integrating these disciplines, agricultural scientists and farmers can significantly improve seed performance, resulting in higher yields and crops that are more resilient. As research continues to advance, the potential for seed breeding to transform agricultural practices increases, promising a more sustainable and productive future for global food production. In summary, seed priming is a valuable tool in modern agriculture, offering benefits that include enhanced germination, better stress tolerance, improved yields, and reduced environmental impact. As agricultural challenges continue to evolve, seed priming stands out as a promising technique to help meet the demands of a growing global population.

José Antonio Hernández Cortés, Gregorio Barba-Espín, and Pedro Diaz-Vivancos

Guest Editors

seeds

Editorial

Seed Priming Technology: Current Perspectives

José Antonio Hernández *, Gregorio Barba-Espín and Pedro Díaz-Vivancos

Group of Fruit Trees Biotechnology, Department of Plant Breeding, CEBAS-CSIC, Campus Universitario de Espinardo, 30100 Murcia, Spain; gbespin@cebas.csic.es (G.B.-E.); pdv@cebas.csic.es (P.D.-V.)
* Correspondence: jahernan@cebas.csic.es

Seeds are vital components in the life cycle of plants, as they are responsible for the propagation of new plant generations. Seed germination and early seedling development are delicate stages in the plant growth cycle, especially when faced with challenging environmental conditions. Additionally, certain plant species require a period of dormancy before restarting growth. Seed dormancy is a state where a seed is unable to germinate, even when provided with optimal conditions for growth [1]. Various techniques have been employed to enhance seed germination and seedling growth under optimal and stress conditions in both orthodox and dormant seeds. Seed priming facilitates germination by controlled hydration and dehydration of seeds prior to sowing, and it has been extensively utilized to boost germination rates and seedling development under various stressors [2–5]. This practice is highly effective in ensuring uniform germination, making it a valuable tool for nurseries. Similarly, different compounds applied during imbibition have proven effective for research purposes and to enhance seed germination and seedling vigor [6].

In that regard, this Special Issue welcomes reviews and experimental research that explore physiological, biochemical, and molecular aspects related to the effect of seed priming on germination and seedling establishment and regarding their interaction with environmental stresses. Seed priming has been proven to improve seed germination in a wide range of plant species and is currently used by seed companies to improve seed germination and plant performance. In an excellent review published in this Special Issue, Corbineau et al. [7] presented extensive information on hydropriming and osmopriming techniques to improve germination, making it faster and more uniform, even in aged seeds, and correlating this with a decline in lipid peroxidation and the increase in reactive oxygen species (ROS)-scavenging enzymes [7]. Hydropriming can be defined as the immersion of seeds in water for a specific time period, whereas osmopriming can be defined as the immersion of seeds in aerated solutions of low water potential (-1.0 to -2.0 MPa), such as polyethylene-glycol, sugars, alcohol, or different salts, including nitrates, phosphates, and chlorides, among others [7]. Currently, omics technologies enable us to gain a deeper insight into the seed priming process as a whole, including the identification of translation initiation factors and genes involved in DNA methylation or acetylation. A better knowledge of the biochemical and molecular mechanisms involved in the effects of priming on seed germination could allow the discovery of molecular and biochemical markers for different priming treatments [7].

Some of the works included in the Special Issue have evaluated the use of potassium nitrate (KNO_3) treatments on germination and early seed growth in the presence or absence of NaCl. The germination of the halophyte *Lobularia maritima* decreased in the presence of 100 and 200 mM NaCl and was even inhibited by 300 mM NaCl. However, the germination rate was improved by the seed priming treatment with 10 mM KNO_3 or 50 mM thiourea, both in the presence and absence of saline treatment. Other seed priming treatments, such as 10 μM salicilyc acid or 1 mM proline, even produced negative effects in the presence of 50 mM NaCl [8]. The positive effect of KNO_3 in seed germination could be dependent on the plant species. In this case, Adhikari et al. [9] described that in the presence of 100 mM NaCl, both 50 mM KNO_3 and 3 mM gibberellic acid (GA_3) treatments had no positive

Citation: Hernández, J.A.; Barba-Espín, G.; Díaz-Vivancos, P. Seed Priming Technology: Current Perspectives. *Seeds* **2024**, *3*, 540–543. https://doi.org/10.3390/seeds3040036

Received: 13 September 2024
Revised: 8 October 2024
Accepted: 10 October 2024
Published: 11 October 2024

effects on the germination of *Lactuca sativa* L. seeds, in terms of germination parameters, morphological traits, or biomass accumulation [9]. However, the hydro-priming of *Lactuca sativa* L. seeds (distilled water soaking for 2 h 30 min) reached a better synchronized germination under NaCl stress but also a higher biomass accumulation, which correlated with the lowest electrolyte leakage (EL) values in relation to the other priming treatments and the non-primed seeds. EL is considered as an oxidative stress parameter that indicates membrane damage [10]. Thus, these results suggest that hydro-primed seeds showed the lowest membrane damage, under salt stress than the other treated seeds.

The effect of KNO_3 in seed germination can also depend on the concentration used for the priming treatment. Hernandez et al. [11] described a positive effect of low KNO_3 levels (0.25 mM) on early pea seedling growth that correlated with a higher water uptake rate during the germination process. However, higher KNO_3 levels (40 mM) reduced the seedling growth. This latter response correlated with a decrease in the activity of H_2O_2-scavenging enzymes, ascorbate peroxidase (APX) and peroxidase (POX), and with the lowest reduced ascorbate contents, accumulating the oxidized form of ascorbate [11]. The effectiveness of KNO_3 in breaking seed dormancy, stimulating seed germination, and ensuring uniform growth in different plant species has also been reported. Nevertheless, the specific mechanism by which KNO_3 enhances seed germination and seedling establishment remains unknown [12]. Studies suggest that plant hormones are essential for the germination process, and it has been shown that KNO_3 can modify the levels of GA and abscisic acid (ABA) during germination [13]. In their study, Hernandez et al. [11] reported that seed priming treatment with low KNO_3 levels (0.25 mM) strongly increased the GA_1 concentration in pea seedlings and reduced the level of ABA in both pea seedlings and cotyledons, leading to a significant decline in the ABA/GAs ratio.

It is well known that salinity impairs seed germination in all plant species, and in addition, soil salinization is an increasing challenge. Other works in this Special Issue have also addressed the use of seed priming treatments to improve seed germination under saline conditions. In this regard, Kataria et al. [14] studied the role of Ca^{2+} and cyclic GMP to mitigate the negative effects of salinity on germination through the magnetopriming of soybean seeds. The authors found that salinity (50 mM NaCl) declined different parameters related to seed germination (early seedling growth, vigor index, total amylase, protease, and nitrate reductase). However, magnetopriming and Ca^{2+} treatment enhanced all of these measured parameters along with a remarkable increase in ROS and nitric oxide (NO) content. The authors concluded that exposure to a magnetic field can activate calcium receptors and target proteins, leading to enhanced calcium and cGMP signaling. This signaling pathway may trigger ROS and NO production, which in turn may activate total amylase, protease, and nitrate reductase activities, accelerating seed germination and promoting seedling growth under salt stress [13]. Salicylic acid (SA) and H_2O_2 were tested during kidney bean germination in the presence of 150 mM NaCl [15]. The authors reported that salinity produced a decrease in different parameters related to seed germination, such as percentage of germination, germination index, seed vigor index, shoot length, root length, and plant growth. The results showed that SA and H_2O_2 treatments, as well as hydropriming, stimulated all the mentioned parameters and enhanced the relative water content and photosynthetic pigments in kidney bean seedlings. The authors concluded that exogenous applications of SA and H_2O_2 can improve salt tolerance and enhance germination and seedling features of kidney beans [15].

The effect of H_2O_2 on seed germination and seedling performance in peach was also investigated by Barba-Espín et al. [16]. The authors found that H_2O_2 imbibition of peach seeds without endocarp after 8 weeks of stratification increased germination rate and resulted in seedlings displaying good vegetative growth. The H_2O_2 imbibition also affected the levels of ascorbate (AsA) and glutathione and produced a decrease in abscisic acid and an increase in jasmonic acid in peach seedlings. Thus, H_2O_2 priming reduced the stratification period needed to germinate peach seeds, which can bring practical application to peach breeding programs and nurseries, as well as to other *Prunus* species [16].

In recent years, seed priming with AsA has been successfully used as a strategy to improve germination and plant productivity, given its crucial role during seed germination. In an excellent review, Terzaghi and De Tullio [17] detailed very interesting aspects of ascorbate metabolism and the dynamic regulation of the AsA system during seed development and seed germination, including the changes detected in the reduced and oxidized forms of ascorbate and the activity of the enzymes APX, monodehydroascorbate reductase (MDHAR), and dehydroascorbate reductase (DHAR). Moreover, changes in the expression of the genes involved in the biosynthesis of AsA, genes encoding for APX, and genes involved in its recycling (different MDHARs and DHARs) were also described [17]. The authors showed evidence that priming seeds with AsA at different concentrations is beneficial to plant growth, development, and productivity, both in the absence or in the presence of some stressors such as NaCl [17].

The role of melatonin as an antioxidant is well known, both in plants and in animals [18]. In that regard, García-Cánovas et al. [19] studied the effects of melatonin on germination and seedling growth under stress conditions, i.e., aging seeds and under polyethylene glycol (PEG)-induced drought stress. Melatonin improved germination in aged seeds at different concentrations, with a maximum effect at 20 μM in rice and 0.05 μM in barley and shorgum. In general, melatonin priming also showed a positive effect on early seedling growth in terms of shoot and root length and biomass accumulation. The PEG-induced drought conditions negatively affected all the germination parameters in sorghum. However, the presence of melatonin reverted and even improved all these germination parameters [19]. The authors concluded that priming seeds with melatonin under low germinability conditions relieves stress and improves both germination and seedling growth, and that this effect could be linked to the antioxidant capacity of this molecule increasing the antioxidant mechanisms, thus controlling the ROS overproduction that takes place under stress conditions. Moreover, the role of melatonin modulating plant hormone contents in tissues should not be ruled out. Consequently, melatonin can act through the modulation of the redox network and plant hormonal responses during the germination process [19].

One of the contributions to this Special Issue involved a transcriptomic analysis using RNAseq technology to explore the molecular mechanisms of osmoprimed *Solanum paniculatum* seeds growing under adverse conditions. The study found that PEG-treated (−1 MPa osmotic potential) seeds had better germination rates under drought stress compared to non-primed seeds, leading to their selection for transcriptomic analysis. Out of 34,640 transcripts, 235 showed differential expression between osmoprimed and water-imbibed seeds, with 232 being upregulated and three being downregulated [20]. The results of differentially expressed transcripts revealed the upregulation of genes encoding proteins related to water, oxidative, saline, and heat stresses as a result of osmopriming, indicating a possible cross-tolerance effect [20].

As a general conclusion, we can state that priming is a commonly employed method to enhance seed quality and facilitate quicker and more uniform seed germination. Currently, priming techniques are widely used at the commercial level. The works published in the Special Issue present information that relates priming techniques with the modulation of antioxidant machinery and the hormone profile (ABA, GAs) in the seeds. In addition, ROS, such as H_2O_2, are able to increase germination in peach through the modulation of AsA, glutathione, and some plant hormones (ABA, JA) levels in peach seedlings. The role of AsA metabolism in seed germination was remarked on in an interesting review as the modulation of ascorbate-related enzymes. The role of antioxidants in seed germination was confirmed by the effect of melatonin on this important process. Furthermore, the Special Issue highlighted that priming can also enhance the germination of low-vigor or old seeds from different species, as well as those subjected to environmental stresses such as salt and drought. Finally, to understand the molecular mechanisms developed by primed seeds to grow under adverse conditions, transcriptomic analysis using RNAseq technology can provide us with very interesting information.

Conflicts of Interest: The authors declare no conflicts of interest.

References

1. Finch-Savage, W.E.; Leubner-Metzger, G. Seed dormancy and the control of germination. *New Phytol.* **2006**, *171*, 501–523. [CrossRef]
2. Paparella, S.; Araújo, S.S.; Rossi, G.; Wijayasinghe, M.; Carbonera, D.; Balestrazzi, A. Seed priming: State of the art and new perspectives. *Plant Cell Rep.* **2015**, *34*, 1281–1293. [CrossRef] [PubMed]
3. Matsushima, K.-I.; Sakagami, J.-I. Effects of Seed Hydropriming on Germination and Seedling Vigor during Emergence of Rice under Different Soil Moisture Conditions. *Am. J. Plant Sci.* **2013**, *4*, 1584–1593. [CrossRef]
4. Forti, C.; Ottobrino, V.; Bassolino, L.; Toppino, L.; Rotino, G.L.; Pagano, A.; Macovei, A.; Balestrazzi, A. Molecular dynamics of pre-germinative metabolism in primed eggplant (*Solanum melongena* L.) seeds. *Hortic. Res.* **2020**, *7*, 87. [CrossRef] [PubMed]
5. Halmer, P. Methods to improve seed performance in the field. In *Handbook of Seed Physiology: Applications to Agriculture*; Benech-Arnold, R.L., Sánchez, R.A., Eds.; Haworth Press: New York, NY, USA, 2004; pp. 125–166.
6. Barba-Espin, G.; Nicolas, E.; Almansa, M.S.; Cantero-Navarro, E.; Albacete, A.; Hernandez, J.A.; Diaz-Vivancos, P. Role of thioproline on seed germination: Interaction ROS-ABA and effects on antioxidative metabolism. *Plant Physiol. Biochem.* **2012**, *59*, 30–36. [CrossRef] [PubMed]
7. Corbineau, F.; Taskiran-Özbingöl, N.; El-Maarouf-Bouteau, H. Improvement of seed quality by priming: Concept and biological basis. *Seeds* **2023**, *2*, 101–115. [CrossRef]
8. Zammali, I.; Dabbous, A.; Youssef, S.; Ben Hamed, K. Effects of chemical priming on the germination of the ornamental halophyte *Lobularia maritima* under NaCl salinity. *Seeds* **2022**, *1*, 99–109. [CrossRef]
9. Adhikari, B.; Olorunwa, O.J.; Barickman, T.C. Seed priming enhances seed germination and morphological traits of *Lactuca sativa* L. under salt stress. *Seeds* **2022**, *1*, 74–86. [CrossRef]
10. Díaz-Vivancos, P.; Rubio, M.; Mesonero, V.; Periago, P.M.; Ros Barceló, A.; Martínez-Gómez, P.; Hernández, J.A. The apoplastic antioxidant system in Prunus: Response to plum pox virus. *J. Exp. Bot.* **2006**, *57*, 3813–3824. [CrossRef] [PubMed]
11. Hernandez, J.A.; Díaz-Vivancos, P.; Acosta-Motos, J.R.; Barba-Espín, G. Potassium nitrate treatment is associated with modulation of seed water uptake, antioxidative metabolism and phytohormone levels of pea seedlings. *Seeds* **2021**, *1*, 5–15. [CrossRef]
12. Bethke, P.C.; Libourel, I.G.L.; Aoyama, N.; Chung, Y.Y.; Still, D.W.; Jones, R.L. The Arabidopsis aleurone layer responds to nitric oxide, gibberellin, and abscisic acid and is sufficient and necessary for seed dormancy. *Plant Physiol.* **2007**, *143*, 1173–1188. [CrossRef] [PubMed]
13. Vidal, A.; Cantabella, D.; Bernal-Vicente, A.; Díaz-Vivancos, P.; Hernández, J.A. Nitrate- and nitric oxide-induced plant growth in pea seedlings is linked to antioxidative metabolism and the ABA/GA balance. *J. Plant Physiol.* **2018**, *230*, 13–20. [CrossRef] [PubMed]
14. Kataria, S.; Shukla, S.; Jumrani, K.; Jain, M.; Gadre, R. Stimulating role of calcium and cyclic GMP in mediating the effect of magnetopriming for alleviation of salt stress in soybean seedlings. *Seeds* **2023**, *2*, 232–245. [CrossRef]
15. Tania, S.S.; Rhaman, M.S.; Rauf, F.; Rahaman, M.d.M.; Kabir, M.H.; Hoque, M.d.A.; Murata, Y. Alleviation of salt-inhibited germination and seedling growth of kidney bean by seed priming and exogenous application of salicylic acid (SA) and hydrogen peroxide (H_2O_2). *Seeds* **2022**, *1*, 87–98. [CrossRef]
16. Barba-Espín, G.; Hernández, J.A.; Martínez-Andújar, C.; Díaz-Vivancos, P. Hydrogen peroxide imbibition following cold stratification promotes seed germination rate and uniformity in peach cv. GF305. *Seeds* **2022**, *1*, 28–35. [CrossRef]
17. Terzaghi, M.; De Tullio, M.C. Ascorbic acid in seeds, priming and beyond. *Seeds* **2023**, *2*, 421–435. [CrossRef]
18. Arnao, M.B. Phytomelatonin: Discovery, Content, and Role in Plants. *Adv. Bot.* **2014**, *2014*, e815769. [CrossRef]
19. Garcia-Cánovas, I.; Giraldo-Acosta, M.; Cano, A.; Arnao, M.B.; Hernández-Ruiz, J. Effect of Melatonin on germination and seedling growth in aging seeds or under drought conditions. *Seeds* **2024**, *3*, 341–356. [CrossRef]
20. Bento da Silva, P.; Afonso Vaz, T.A.; Acencio, M.L.; Hilhorst, H.W.M.; Amaral da Silva, E.A. Can osmopriming induce cross-tolerance for abiotic stresses in *Solanum paniculatum* L. seeds? A transcriptome analysis point of view. *Seeds* **2023**, *2*, 382–393. [CrossRef]

seeds

MDPI

Article

Potassium Nitrate Treatment Is Associated with Modulation of Seed Water Uptake, Antioxidative Metabolism and Phytohormone Levels of Pea Seedlings

José A. Hernández *, Pedro Díaz-Vivancos *, José Ramón Acosta-Motos and Gregorio Barba-Espín *

Group of Fruit Trees Biotechnology, Department of Plant Breeding, CEBAS-CSIC, Campus Universitario de Espinardo, 30100 Murcia, Spain; jacosta@cebas.csic.es
* Correspondence: jahernan@cebas.csic.es (J.A.H.); pdv@cebas.csic.es (P.D.-V.); gbespin@cebas.csic.es (G.B.-E.)

Abstract: (1) Background: Seed treatment with potassium nitrate (KNO_3) has been associated with dormancy breaking, improved germination and enhanced seedling growth and uniformity in a variety of plant species. However, the KNO_3 effect seems to be dependent on plant species and treatment conditions. (2) Methods: We describe the effect of incubation of dry pea seeds with different KNO_3 concentration on water uptake kinetic, early seedling growth, antioxidant metabolism and hormone profile in pea seedlings. (3) Results: Low (0.25 mM) KNO_3 levels increased seedling water uptake and growth, whereas high (40 mM) levels decreased seedling growth. KNO_3 treatment differentially affected the antioxidant defences. Low KNO_3 levels maintained the activity of antioxidant enzymes, while high levels reduced the activity of H_2O_2-scavenging enzymes. KNO_3 induced a progressive decline in ascorbate levels and reduced (GSH) and oxidised (GSSG) glutathione. Low KNO_3 levels strongly increased GA_1 and decreased ABA in both seedlings and cotyledons, resulting in a decline in the ABA/GAs ratio. (4) Conclusions: Pea seed treatment with a low KNO_3 level promoted early seedling growth. In this process, an interaction among KNO_3, antioxidant defences and ABA/GAs ratio is proposed.

Keywords: ABA; antioxidant defences; GAs; nitrate; seed germination; seedling growth; water uptake

Citation: Hernández, J.A.; Díaz-Vivancos, P.; Acosta-Motos, J.R.; Barba-Espín, G. Potassium Nitrate Treatment Is Associated with Modulation of Seed Water Uptake, Antioxidative Metabolism and Phytohormone Levels of Pea Seedlings. *Seeds* **2021**, *1*, 5–15. https://doi.org/10.3390/seeds1010002

Academic Editor: Alma Balestrazzi

Received: 7 October 2021
Accepted: 30 November 2021
Published: 2 December 2021

1. Introduction

Seed germination is the most critical stage in crop establishment, determining crop production [1]. Numerous methods have been used to promote seed germination and seedling establishment under normal and stressful conditions. In this sense, seed priming, defined as a pre-sowing treatment which involves controlled hydration of seeds during the first stage of germination, has been widely applied to improve the germination rate and seedling growth under different stress conditions [2–5]. Likewise, seed chemical treatment during imbibition has been successfully applied for both fundamental research purposes and for the stimulation of seed germination and seedling vigour [6–8].

Potassium nitrate (KNO_3) has been demonstrated to break seed dormancy, promote seed germination and enhance growth uniformity in a variety of plant species, including tomato, maize, Arabidopsis and pea [9–12]. However, the mechanism by which KNO_3 improves seed germination and seedling establishment remains unclear. The priming effect of KNO_3 seems to be dependent on the KNO_3 concentration and application method. For example, improved seedling establishment was observed when seeds of *Paspalum vaginatum* (cv. Sea-Spray) were imbibed in 20–50 mM KNO_3 for three days [13]. Conversely, other authors reported that water-imbibed pea seeds (*Pisum sativum* cv. Lincoln) displayed reduced germination and seedling growth when incubated for three days in the presence of 30 or 40 mM KNO_3, whereas 10 mM KNO_3 stimulated early seedling growth. However, when the pea seeds were imbibed with KNO_3 (40 mM) for 24 h and then incubated in distilled water, seedling fresh weight was enhanced, whereas germination rate remained unchanged [12]. Other authors [14] suggested that nitrates may enhance seed germination and early seedling growth due to the dual role of N as a plant essential element for growth and as a signalling molecule.

The effects of KNO_3 on plant hormones' regulation are unclear, with most studies focusing on abscisic acid (ABA) levels. KNO_3 has been reported to positively affect seed germination by modulating ABA metabolism or ABA signalling in developing seeds [15,16]. Gibberellins (GAs) are also important plant hormones for numerous physiological plant processes, including seed germination [17]. Other authors [18] reported that activation of ABA catabolism and GAs biosynthesis is required for seed germination. Low nitrate concentration decreases ABA content, leading to the induction of the *CYP707A2* gene, which encodes an ABA 8′-hydroxylase involved in ABA catabolism [15,19]. In addition, the up-regulation of the *CYP707A2* gene precedes the induction of the *GA3ox2* gene, related to GA biosynthesis [20]. Vidal et al. [12] described that exogenous KNO_3 enhanced GA_4 content and reduced ABA levels, resulting in a decrease in the ABA/GAs ratio. This effect was reversed by the action of the nitric oxide (NO)-scavenger 2-4-carboxyphenyl-4, 4, 5, 5-tetramethylimidazoline-1-oxyl-3-oxide (cPTIO), suggesting an interplay between KNO_3 and NO metabolism [21]. This in turn would indicate the role of NO in seed germination and seed dormancy breaking [9,22].

In addition, exogenous KNO_3 has been reported to lead to an increase in the expression of genes involved in N and C metabolism, as well as in energy production [15]. A role for nitrate reductase induction, related to N assimilation, and the antioxidative metabolism has also been described [11,12]. In that regard, the application of KNO_3 has been reported to increase antioxidant enzymes activity (superoxide dismutase (SOD), catalase (CAT), peroxidase (POX) and ascorbate oxidase (AOX)) in seedlings from different plant species [11,12]. During imbibition, seeds take an increasing amount of oxygen, causing the accumulation of reactive oxygen species (ROS) and a shift in the redox state [23]. Although ROS were originally considered as toxic by-products, different studies revealed that they are used by most organisms as key signalling molecules [24]. In addition, it was recently described that a basal level of ROS is required to support life [25]. In fact, it was recently suggested that the type of oxidative modification integrated into different oxidative signalling pathways regulates many crucial aspects of plant biology [26]. Several works have described the implication of ROS and antioxidant metabolism in the germination process. The scientific literature contains plenty of information concerning the beneficial effects of ROS on germination and seedling growth processes [6–8,27–31]. In these works, the authors showed that a controlled ROS generation during seed imbibition may have a signalling function during germination as well as during the dormancy release process. In contrast, uncontrolled ROS accumulation could delay or even inhibit seed germination [31]. In addition, an interplay ROS/plant hormone has been described during the seed germination process [6–8,28,29].

In the present work, we attempted to explain if direct nitrate addition during imbibition may promote pea seed germination and seedling establishment and vigour. For this purpose, we treated dry pea seeds with different KNO_3 concentrations to study its effect on water uptake kinetic, early seedling growth and antioxidant metabolism-related enzyme activity in pea seedlings. In addition, the effects of KNO_3 on ABA and GAs levels in pea seedlings and cotyledons were also addressed.

2. Materials and Methods

2.1. Plant Material, Culture Conditions, Growth Measurements and Sampling

Pea (*Pisum sativum* cv. Lincoln) seeds were obtained from Ramiro Arnedo S.A, Murcia, Spain. In the first experiment, individual dry seeds were placed inside 3.5 cm diameter plastic cups, onto two discs of filtered paper moistened with 1 mL KNO_3 (0, 0.25, 0.5, 1, 5, 10, 20, 30, 40 or 80 mM). Twenty-five seeds per treatment were arranged in trays and placed inside plastic bags, containing some small holes, to avoid water evaporation, and incubated in darkness at 25 °C in an incubator (MIR-153, Sanyo, Osaka, Japan). The germination percentage and water absorption rate (µL H_2O/g dry weight (DW), calculated as the difference between the final and the initial weights, divided by the initial weight) were registered daily during KNO_3 treatment from days 0 to 4.

Subsequently, and based on the results of the preliminary assays, dry seeds were placed on 15 cm diameter Petri dishes, onto two discs of filtered paper moistened with 7 mL KNO_3 (0, 0.25 or 40 mM). Three to six Petri dishes per treatment containing 20 seeds per plate (biological replicate) were arranged and incubated at 25 °C for four days in an incubator (MIR 153, Sanyo) in darkness. Subsequently, seedling growth (length and mass) was recorded. Seedlings were separated from cotyledons and both were frozen in liquid nitrogen and stored at −80 °C for further analysis.

2.2. Enzyme Extraction and Assays

All operations were carried out at 4 °C. Four-day-old pea seedlings were homogenised and prepared for enzymatic analyses as described [12]. The activity of the antioxidant enzymes (ascorbate (ASC)-glutathione (GSH) cycle enzymes, SOD, POX, CAT and ascorbate oxidase (AOX)) was measured as previously reported [6,32]. The protein concentration was calculated according to Bradford [33]. The analyses were performed in a UV/Vis V-630 Bio spectrophotometer (Jasco, Tokyo, Japan).

2.3. Ascorbate and Glutathione Analyses

Frozen pea seedlings were ground into a fine powder in the presence of liquid nitrogen. Then, a 1M $HClO_4$ solution containing 1 mM EDTA and 1% (*w/v*) polyvinylpyrrolidone phosphate was added (1/3, *w/v*). The resulting extract was centrifuged at 12,000× *g* for 10 min at 4 °C. The pH of the supernatant was adjusted to 5.5–6 with 5 M K_2CO_3. The new mixture was centrifuged at 12,000× *g* for 1 min to eliminate the precipitate of $KClO_4$ formed. The resulting supernatant was used to quantify the oxidised and reduced ascorbate and glutathione forms [32].

2.4. Analysis of Plant Hormones

The analysis of ABA and GAs was performed in seedlings and cotyledons. Frozen samples were first lyophilised and then ground into a fine powder in a mortar. One 30-mg aliquot of each sample was sent to the Plant Hormones Quantification Service of the IBMCP (CSIC, Valencia) and quantified as described [34].

2.5. Statistical Analyses

Analyses of germination percentage, water uptake and seedling growth were performed on at least 20 biological replicates (individual seedlings). The remaining analysis was on three to six biological replicates (the pool of seedlings from a plate). Data were expressed as the mean ± SE. The data were analysed by one-way ANOVA followed by a Tukey's multiple range test ($p \leq 0.05$) using the SPSS 26 software (IBM SPSS Statistics, Chicago, IL, USA). All experiments were repeated independently at least twice with similar results.

3. Results

In a preliminary experiment, we assayed the effect of different KNO_3 concentrations on the water uptake rate and germination rate of pea seeds. In general, higher KNO_3 concentrations (10–80 mM) delayed seed germination, although at days 3 and 4 the germination rate was similar to that of control seeds (Supplemental Table S1a). Lower KNO_3 concentrations (0.25–5 mM) showed similar germination rates when compared to the control treatment at every day of treatment (Supplemental Table S1b). Taking into account that the highest seedling fresh and dry weights were registered with 0.25 mM KNO_3, this concentration, together with 40 mM KNO_3 as a representative of a high concentration, were used for further experiments.

Figure 1 shows the seedling water uptake for the control, 0.25 mM and 40 mM KNO_3. The 40 mM KNO_3 treatment did not alter seedling water uptake compared to control seeds, whereas 0.25 mM KNO_3 increased it significantly with respect to the control at days 1 and 3.

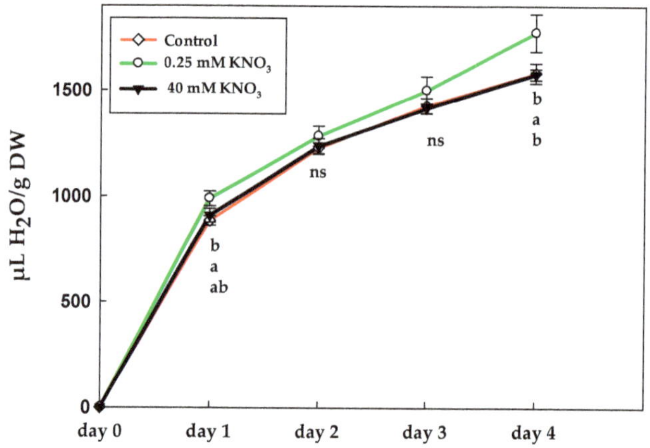

Figure 1. Effect of KNO_3 on the rate of water uptake of pea seedlings during 4 days of incubation. Each seed was weighed before being subjected to any treatment (initial dry weight (DW), day 0), and was then deposited in the bottom of a 3.5 cm diameter plastic cup, onto two discs of filter paper moistened with 0 (control), 0.25 and 40 mM KNO_3. The seeds were weighed daily and the rate of water absorption calculated as μL H_2O/g DW. Different letters (ordered from the top to the bottom for control, 0.25 mM and 40 mM KNO_3, respectively) within the same day indicate significant differences according to Tukey's test ($p \leq 0.05$).

A different effect for 0.25 and 40 mM KNO_3 was observed in the plant growth parameters (Figure 2). In general, in pea seedlings, the low KNO_3 concentration increased both FW and length by more than 15%, whereas 40 mM KNO_3 decreased the length and FW by 15% and 25%, respectively (Figure 2).

According to the ANOVA analysis, the KNO_3 treatments significantly affected all the enzymes of the ASC-GSH cycle as well as POX activity, measured at day 4 of treatment. At 40 mM KNO_3, a near 2-fold decrease in APX and POX activity was observed (Table 1). Conversely, with the same treatment, DHAR and GR significantly increased by about 60%. An increase in MDHAR, DHAR and GR (ascorbate and glutathione recycling activities) was also observed for the 0.25 mM KNO_3 treatment (Table 1). Moreover, POX activity significantly increased (near 2.5-fold) in 0.25 mM KNO_3 with respect to the 40 mM KNO_3 treatment, though the increase with respect to the control (27%) was non-significant (Table 1).

Table 1. Effect of KNO_3 on the activity of different antioxidant enzymes on four-day-old pea seedlings. Ascorbate oxidase (AOX), ascorbate peroxidase (APX), monodehydreascorbate reductase (MDHAR), dehydreascorbate reductase (DHAR), glutathione reductase (GR) and peroxidase (POX) activity are expressed in nmol $\times$ min^{-1} $\times$ mg^{-1} protein. Superoxide dismutase (SOD) activity is expressed in Units (U) $\times$ mg^{-1} protein.

KNO_3 (mM)	AOX	APX	MDHAR	DHAR	GR	POX	SOD
0	60.5 ± 4.1	39.9 ± 4.2 a	198 ± 12.9 b	0.52 ± 0.05 b	50.8 ± 3.0 b	1383 ± 172 ab	34.8 ± 2.5
0.25	51.9 ± 1.7	34.8 ± 2.2 a	316 ± 15.9 a	0.80 ± 0.08 a	85.8 ± 3.1 a	1765 ± 243 a	37.5 ± 3.8
40	53.1 ± 3.8	18.2 ± 2.2 b	165 ± 19.9 b	0.85 ± 0.04 a	80.6 ± 3.9 a	713 ± 43 b	27.7 ± 2.8
[a] F	2.39 ns	13.7 **	23.4 ***	23.4 ***	31.7 ***	8.46 **	2.71 ns

[a] Significant F values from one-way ANOVA are denoted at 99.9% (***) or 99% (**) levels of probability (ns, not significant). Data represent the mean ± SE from at least six measurements. Different letters indicate significant differences according to Tukey's test ($p \leq 0.05$).

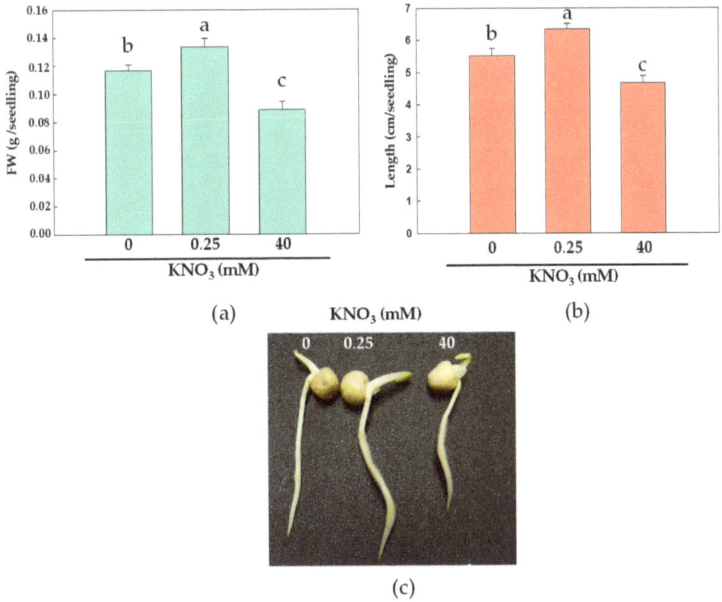

Figure 2. Effect of KNO_3 on the (**a**) fresh weight (FW) and (**b**) length of pea seedlings after four days of incubation. (**c**) Picture showing the visual effects of KNO_3 on seedling growth. Data represent the mean ± SE from at least 30 measurements. Different letters indicate significant differences according to Tukey's test ($p \leq 0.05$).

The KNO_3 treatments also affected the ascorbate and glutathione content (Table 2). In that sense, KNO_3 strongly decreased the reduced ascorbate (ASC) content in a concentration-dependent manner (Table 2). The oxidised ascorbate (DHA) was only detected in those samples treated with the highest KNO_3 level (Table 2). At 0.25 mM KNO_3, a decrease in both reduced (GSH) and oxidised (GSSG) glutathione levels, and thus in the total concentration of glutathione, was observed. However, an increase in the redox state of the glutathione pool was observed (Table 2). In contrast, at 40 mM KNO_3, GSH and total glutathione levels were not affected, though a decrease in GSSG was observed and, consequently, an increase in the redox state of the glutathione pool was registered (Table 2).

Table 2. Effect of KNO_3 on ascorbate and glutathione concentration in pea seedlings. Reduced (ASC) and oxidised (DHA) ascorbate and reduced (GSH), oxidised (GSSG) and total glutathione are expressed in nmol g^{-1} fresh weight (FW). (nd: not detected.)

KNO_3 (mM)	ASC	DHA	GSH	GSSG	GSH/ GSH + GSSG	Total Glutathione
0	1401 ± 73 a	nd	390 ± 19 a	20.16 ± 3.12 a	0.951	425 ± 32 a
0.25	535 ± 34 b	nd	263 ± 14 b	9.27 ± 2.06 c	0.966	281 ± 11 b
40	375 ± 85 b	7.60 ± 4.39	447 ± 26 a	14.10 ± 1.81 b	0.969	452 ± 29 a
[a] F	60.9 ***		32.9 ***	21.7 ***		15.2 ***

[a] Significant F values from one-way ANOVA are denoted at 99.9% (***) levels of probability. Data represent the mean ± SE from at least six measurements. Different letters indicate significant differences according to Tukey's test ($p \leq 0.05$).

Potassium nitrate affected the ABA and GAs levels in both the seedlings and the cotyledons. GA_4 was detected in both tissues, while GA_1 was only detected in the seedlings. In seedlings treated with 0.25 mM KNO_3, a 6-fold increase in GA_1 as well as a slight decrease in GA_4 was observed (Figure 3a). In cotyledons, the effect of KNO_3 on GA_4 levels was not statistically significant (Figure 3b).

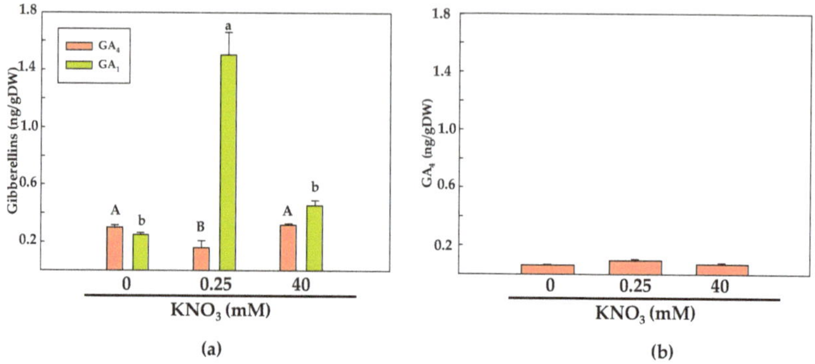

Figure 3. Effect of KNO_3 on gibberellins (GAs) levels in (**a**) pea seedlings and (**b**) cotyledons after four days of incubation. Data represent the mean ± SE from at least three measurements. Different letters (uppercase and lowercase letters for GA_4 and GA_1, respectively) indicate significant differences according to Tukey's test ($p \leq 0.05$). (DW, dry weight.)

The effect on ABA content varied depending on KNO_3 concentration and the type of tissue (seedling or cotyledon). In seedlings, at 0.25 mM KNO_3, a 54% decrease in the ABA level was observed, while 40 mM KNO_3 concentration produced a 1.7-fold increase in ABA, compared to control seedlings (Figure 4a). ABA levels in the cotyledons were six times lower than in seedlings under control conditions. At 0.25 mM KNO_3, ABA levels declined in the cotyledon, though no effect was observed at 40 mM KNO_3 (Figure 4a). As a consequence of the KNO_3-induced changes in ABA and GAs in pea seedlings and cotyledons, an important decrease in the ABA/total GAs ratio occurred in the cotyledons and, especially, in the seedlings (Figure 4b).

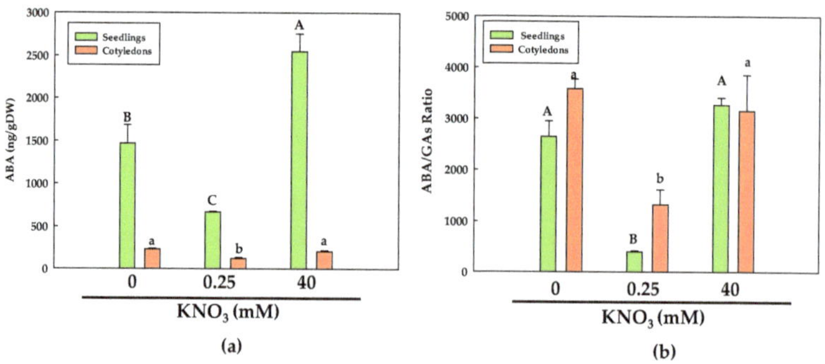

Figure 4. Effect of KNO_3 on (**a**) abscisic acid (ABA) levels and (**b**) ABA/GAs ratio in pea seedlings and cotyledons after four days of incubation. Data represent the mean ± SE from at least three measurements. Different letters (uppercase and lowercase letters for seedlings and cotyledons, respectively) indicate significant differences according to Tukey's test ($p \leq 0.05$). (DW, dry weight).

4. Discussion

Nitrogen is a macronutrient that may be a limiting factor for plant growth. At low concentrations, nitrate can stimulate seed germination in a variety of different plant species [14]. The mechanism of action of KNO_3 on the improvement of seed germination and/or early growth is far from being completely understood. The complexity about nitrate effects on seed germination and early seedling growth could be due to its dual role as a nutrient and a signalling molecule [14]. Nitrate stimulation of seed germination is often associated with plant species whose seeds require light for germination [35,36].

Potassium nitrate has been reported to improve pea seed germination and plant performance, but this effect is dependent on the concentration and the mode of application [1–4].

In the present work, we applied KNO_3 directly to dry seeds; under these conditions a low KNO_3 concentration, such as 0.25 mM, increased the early seedling growth of peas, but there was no effect on the germination rate. This effect was parallel with an increased water uptake by the seed. In that regard, water uptake is an integral requirement for the initiation and completion of the germination process [37]. In contrast, when 40 mM KNO_3 was applied directly to the dry seed, there was a significance decrease in water uptake. However, when the application of low KNO_3 followed imbibition in distilled H_2O for 24 h, a contrasting effect on seedling growth was observed [12]. A minor effect was found at a low concentration (1 mM), whereas increased seedling FW and length were observed at a high concentration (30 mM). On the other hand, direct imbibition in 30 mM KNO_3 reduced both the fresh mass and length of seedlings [12].

4.1. Antioxidant Metabolism

There is limited information on the effect of KNO_3 priming on the antioxidant metabolism of plant seedlings. Only a few papers have reported some connection with POX, SOD, CAT and AOX enzymes [11,12]. In the present study, pea seedlings were found to contain very low DHAR activity, suggesting they mainly use MDHAR activity for ascorbate recycling, which utilises NADH as an electron donor. From an energy point of view, this is much more efficient than the DHAR pathway, which uses GSH as a source of reducing power [38]. Similarly, using the same pea cultivar, a higher MDHAR activity than that of DHAR was also reported at subcellular level [39,40]. Moreover, in pea seeds (cv. Alaska) imbibed in 20 mM H_2O_2, no DHAR activity was recorded, leading to a DHA accumulation in pea seedlings [6]. In addition, the seedlings subjected to 0.25 mM KNO_3 treatment displayed 2-fold higher MDHAR activity than seedlings treated with 40 mM KNO_3, suggesting a higher capacity to recycle ascorbate. A similar response was observed with APX and POX activity, indicating a reduced ability to control H_2O_2 levels in 40 mM KNO_3-treated seedlings when compared to controls and seedlings treated with a low KNO_3 concentration. MDHAR activity showed a different response to KNO_3 treatments in comparison to that which occurred with APX and POX activity. In this sense, low KNO_3 treatment increased both MDHAR activity and the H_2O_2-scavenging enzymes APX and POX, whereas high KNO_3 treatment produced a decrease in the activity of these enzymatic antioxidants. Both APX and some type of POXs can use ASC as an electron donor. These types of POXs can thus oxidise ASC and organic phenols at comparable rates [41]. Therefore, if APX and POX decreased, a lower level of ASC is oxidised and thus lower MDHA can be generated. This could explain the decline in MDHAR recorded by 40 mM KNO_3 treatment.

Similarly, H_2O_2-primed pea seeds also showed an increase in APX and POX activity [6]. It may be that pea seedlings treated with 0.25 mM KNO_3 displayed an efficient regulation of ROS generation when compared to those treated with 40 mM KNO_3, which in turn correlated with an enhanced seedling growth. The increase of POX and the maintenance of APX activity by KNO_3 treatment were also reported by Vidal et al. [12]. In tomato seeds, KNO_3 increased germination rates and enhanced SOD and catalase activities but had no effect on APX activity [11].

Information about the effect of KNO_3 treatments on the ascorbate pool is very scarce. Under the assayed conditions, KNO_3 progressively decreased ASC content. However, DHA only accumulated in seedlings treated with 40 mM KNO_3. According to our results, this response suggests that an effect on the biosynthetic pathway of ascorbate could take place. These results are in contrast to those reported by Vidal et al. [12], who observed a 3-fold increase in ASC in response to 10 mM KNO_3 or 50 μM sodium nitroprusside (SNP, a NO-donor). This may reflect the differences in the experimental procedures between the present study and the one reported previously [12]. Regarding glutathione levels, 0.25 mM KNO_3 produced a decrease in both GSH and GSSG, leading to a slight increase in the redox state of the glutathione pool. Likewise, 20 mM H_2O_2-treated pea seeds showed a decline in GSH and GSSG [6]. In this study, 40 mM KNO_3 had no effect on GSH but reduced GSSG levels, thus increasing the redox state of the glutathione pool. The reduced accumulation in GSSG in both KNO_3 treatments can be linked to the observed increase in GR activity. Again, these results are in contrast to those reported by Vidal et al. [12], where no significant differences in GSH content with 10 mM KNO_3 were reported. In addition, these authors reported an accumulation of GSSG, which resulted in a decrease in the redox state of the glutathione pool. This highlights that the effect of KNO_3 treatment is highly dependent on the mode of application. Alternatively, decreased ASC and GSH levels in seed treated with 0.25 mM KNO_3 can be related to altered metabolism rather than reduced recycling of both molecules. In this sense, MDHAR, as well as DHAR and GR, activity was nearly 60% higher than in the control seedlings.

4.2. Plant Hormones

Potassium nitrate has been reported to affect GAs and ABA metabolism, though there is limited information available in this regard. It has been reported that exogenous KNO_3 or NO accumulation can modulate ABA and GAs content by increasing ABA catabolism as well as the up-regulation of GAs biosynthesis, respectively [15,18,20]. Vidal et al. [12] observed that KNO_3 or SNP treatments increased GAs levels and decreased ABA concentration in pea seedlings. These effects were reversed by incubation with cPTIO (a NO-scavenger), suggesting that, in part, the effect of KNO_3 in seedling growth and plant hormone levels may be due to partial generation of NO from KNO_3 [21]. In that sense, NO has a significant role in seed germination and the breaking of seed dormancy [9,22,27].

It is well known that GAs, together with ABA, are involved in seed dormancy and germination, promoting germination in many plant species [21,42]. In this study, a major proportion of GAs were present in the seedlings (GA_1 and GA_4), with only a small amount of GA_4 reported in the cotyledon. This suggests that during the growth and development of the seedling, GAs could be mobilised from the cotyledon to the rest of the plant, although an increase in GAs biosynthesis cannot be ruled out [20]. Hormone levels in particular organelles would seem to be dependent on metabolism and transportation [21]. GAs biosynthesis genes are expressed in different tissues in the embryonic axes of Arabidopsis seeds [42]. Therefore, it is likely that GAs and their precursors are actively transported inside the embryo and/or from the cotyledon to the embryo by specific transporters.

Decreased ABA levels in seedlings treated with 0.25 mM KNO_3 were correlated with enhanced seedling biomass. In addition, seeds treated with 40 mM KNO_3, which achieved the lowest seedling growth, also had the highest ABA levels in their seedlings. During seed dormancy breaking and germination in Arabidopsis, both a decline in ABA content and an accumulation of NO was reported [18]. This response was parallel to the induction of the *ABA-8-hydroxylase* gene, involved in ABA catabolism [15,18]. A decline in ABA levels has also been described in H_2O_2-treated pea seeds [6,8,43]. These authors suggested that this response can result from either the stimulation of ABA catabolism or the inhibition or slowdown of its biosynthesis.

As a result of the effect of KNO_3 on GAs and ABA content, a decrease in the ABA/GAs ratio was observed by 0.25 mM KNO_3, mainly in the seedlings, when compared to controls and samples treated with 40 mM KNO_3. It has been demonstrated that the ABA/GAs balance is crucial during the early stages of germination [44]. It has also been suggested that the key role of phytohormones such as ABA and GA during seed germination is interdependent with ROS metabolism [27]. A tight control of ROS production appears to be crucial for seed germination [45]. In fact, an accumulation of ROS and NO in germinating seeds, as well as an enhanced seed germination by the exogenous application of H_2O_2, has been reported [6,32,43]. Thus, KNO_3 can trigger a higher seedling growth by regulating ABA and GAs metabolisms, likely associated with an over-generation of NO from KNO_3.

5. Conclusions

The application of a low KNO_3 concentration to dry pea seeds promoted early seedling growth, which was linked to the maintenance (APX, SOD, POX) and/or the increase

(MDHAR, DHAR, GR) of antioxidant defences, leading to an efficient regulation of the ROS generation. In addition, and although 0.25 mM KNO_3 decreased GSH levels, a slight increase in the redox state of glutathione pool, which could be associated with an increase in GR, was observed. In general, a low KNO_3 concentration was associated with an increased GA_1 and decreased ABA in seedlings, which resulted in a decline in the ABA/GAs ratio. Data also suggest a modulation of GAs and ABA metabolism by KNO_3, in which a partial role of NO could not be ruled out. Furthermore, we suggest an interaction among KNO_3, antioxidant defences and the modulation of the ABA/GA ratio during the early growth of pea seedlings (Figure 5).

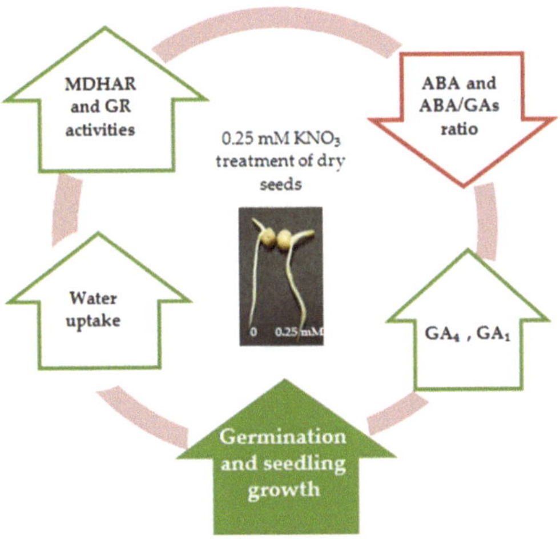

Figure 5. Simplified model summarising the effect of low KNO_3 concentration on promoting pea seed germination and seedling growth.

Supplementary Materials: The following are available online at https://www.mdpi.com/article/10.3390/seeds1010002/s1, Table S1a: Effect of different KNO_3 treatments on the germination rate of pea seeds, Table S1b: Effect of different KNO_3 concentrations on some growth parameters in pea seedling after 4 days of treatments.

Author Contributions: Conceptualization, J.A.H. and P.D.-V.; methodology, J.A.H. and P.D.-V.; validation, J.A.H., P.D.-V., G.B.-E. and J.R.A.-M.; formal analysis, J.A.H. and J.R.A.-M.; investigation, J.A.H. and P.D.-V.; resources, J.A.H. and P.D.-V.; data curation, J.A.H., P.D.-V., G.B.-E. and J.R.A.-M.; writing—original draft preparation, J.A.H. and P.D.-V.; writing—review and editing, J.A.H., P.D.-V., G.B.-E. and J.R.A.-M. All authors have read and agreed to the published version of the manuscript.

Funding: This research received no external funding.

Institutional Review Board Statement: Not applicable.

Informed Consent Statement: Not applicable.

Data Availability Statement: The study did not report any data.

Acknowledgments: We thank Isabel Lopez-Diaz and Esther Carrera for the hormone quantification carried out at the Plant Hormone Quantification Service, IBMCP, Valencia, Spain.

Conflicts of Interest: The authors declare no conflict of interest.

References

1. Rajjou, L.; Duval, M.; Gallardo, K.; Catusse, J.; Bally, J.; Job, C.; Job, D. Seed germination and vigor. *Annu. Rev. Plant Biol.* **2012**, *63*, 507–533. [CrossRef]
2. Paparella, S.; Araújo, S.S.; Rossi, G.; Wijayasinghe, M.; Carbonera, D.; Balestrazzi, A. Seed priming: State of the art and new perspectives. *Plant Cell Rep.* **2015**, *34*, 1281–1293. [CrossRef]

3. Matsushima, K.-I.; Sakagami, J.-I. Effects of Seed Hydropriming on Germination and Seedling Vigor during Emergence of Rice under Different Soil Moisture Conditions. *Am. J. Plant Sci.* **2013**, *4*, 1584–1593. [CrossRef]
4. Forti, C.; Ottobrino, V.; Bassolino, L.; Toppino, L.; Rotino, G.L.; Pagano, A.; Macovei, A.; Balestrazzi, A. Molecular dynamics of pre-germinative metabolism in primed eggplant (*Solanum melongena* L.) seeds. *Hortic. Res.* **2020**, *7*, 87. [CrossRef]
5. Ren, Y.; Wang, W.; He, J.; Zhang, L.; Wei, Y.; Yang, M. Nitric oxide alleviates salt stress in seed germination and early seedling growth of pakchoi (*Brassica chinensis* L.) by enhancing physiological and biochemical parameters. *Ecotoxicol. Environ. Saf.* **2020**, *187*, 109785. [CrossRef] [PubMed]
6. Barba-Espin, G.; Diaz-Vivancos, P.; Clemente-Moreno, M.J.; Albacete, A.; Faize, L.; Faize, M.; Pérez-Alfocea, F.; Hernández, J.A. Interaction between hydrogen peroxide and plant hormones during germination and the early growth of pea seedlings. *Plant Cell Environ.* **2010**, *33*, 981–994. [CrossRef]
7. Barba-Espín, G.; Diaz-Vivancos, P.; Job, D.; Belghazi, M.; Job, C.; Hernández, J.A. Understanding the role of H_2O_2 during pea seed germination: A combined proteomic and hormone profiling approach. *Plant Cell Environ.* **2011**, *34*, 1907–1919. [CrossRef]
8. Barba-Espin, G.; Nicolas, E.; Almansa, M.S.; Cantero-Navarro, E.; Albacete, A.; Hernandez, J.A.; Diaz-Vivancos, P. Role of thioproline on seed germination: Interaction ROS-ABA and effects on antioxidative metabolism. *Plant Physiol. Biochem.* **2012**, *59*, 30–36. [CrossRef] [PubMed]
9. Bethke, P.C.; Libourel, I.G.L.; Jones, R.L. Nitric oxide reduces seed dormancy in Arabidopsis. *J. Exp. Bot.* **2006**, *57*, 517–526. [CrossRef]
10. Anosheh, H.P.; Sadeghi, H.; Emam, Y. Chemical priming with urea and KNO_3 enhances maize hybrids (*Zea mays* L.) seed viability under abiotic stress. *J. Crop Sci. Biotechnol.* **2011**, *14*, 289–295. [CrossRef]
11. Lara, T.S.; Lira, J.M.S.; Rodrigues, A.C.; Rakocevic, M.; Alvarenga, A.A. Potassium Nitrate Priming Affects the Activity of Nitrate Reductase and Antioxidant Enzymes in Tomato Germination. *J. Agric. Sci.* **2014**, *6*, 72. [CrossRef]
12. Vidal, A.; Cantabella, D.; Bernal-Vicente, A.; Díaz-Vivancos, P.; Hernández, J.A. Nitrate- and nitric oxide-induced plant growth in pea seedlings is linked to antioxidative metabolism and the ABA/GA balance. *J. Plant Physiol.* **2018**, *230*, 13–20. [CrossRef]
13. Shim, S.I.; Moon, J.-C.; Jang, C.S.; Raymer, P.; Kim, W. Effect of Potassium Nitrate Priming on Seed Germination of Seashore Paspalum. *HortScience* **2008**, *43*, 2259–2262. [CrossRef]
14. Duermeyer, L.; Khodapanahi, E.; Yan, D.; Krapp, A.; Rothstein, S.J.; Nambara, E. Regulation of seed dormancy and germination by nitrate. *Seed Sci. Res.* **2018**, *28*, 150–157. [CrossRef]
15. Matakiadis, T.; Alboresi, A.; Jikumaru, Y.; Tatematsu, K.; Pichon, O.; Renou, J.P.; Kamiya, Y.; Nambara, E.; Truong, H.N. The Arabidopsis abscisic acid catabolic gene CYP707A2 plays a key role in nitrate control of seed dormancy. *Plant Physiol.* **2009**, *149*, 949–960. [CrossRef] [PubMed]
16. Chahtane, H.; Kim, W.; Lopez-Molina, L. Primary seed dormancy: A temporally multilayered riddle waiting to be unlocked. *J. Exp. Bot.* **2017**, *68*, 857–869. [CrossRef] [PubMed]
17. Ogawa, K.; Iwabuchi, M. A mechanism for promoting the germination of Zinnia elegans seeds by hydrogen peroxide. *Plant Cell Physiol.* **2001**, *42*, 286–291. [CrossRef] [PubMed]
18. Liu, Y.; Ye, N.; Liu, R.; Chen, M.; Zhang, J. H_2O_2 mediates the regulation of ABA catabolism and GA biosynthesis in Arabidopsis seed dormancy and germination. *J. Exp. Bot.* **2010**, *61*, 2979–2990. [CrossRef]
19. Okamoto, M.; Kuwahara, A.; Seo, M.; Kushiro, T.; Asami, T.; Hirai, N.; Kamiya, Y.; Koshiba, T.; Nambara, E. CYP707A1 and CYP707A2, which encode abscisic acid 8′-hydroxylases, are indispensable for proper control of seed dormancy and germination in Arabidopsis. *Plant Physiol.* **2006**, *141*, 97–107. [CrossRef]
20. Ogawa, M.; Hanada, A.; Yamauchi, Y.; Kuwahara, A.; Kamiya, Y.; Yamaguchi, S. Gibberellin biosynthesis and response during Arabidopsis seed germination. *Plant Cell* **2003**, *15*, 1591–1604. [CrossRef] [PubMed]
21. Nonogaki, H. Seed biology updates—Highlights and new discoveries in seed dormancy and germination research. *Front. Plant Sci.* **2017**, *8*, 524. [CrossRef]
22. Beligni, M.V.; Lamattina, L. Nitric oxide stimulates seed germination and de-etiolation, and inhibits hypocotyl elongation, three light-inducible responses in plants. *Planta* **2000**, *210*, 215–221. [CrossRef] [PubMed]
23. Rosental, L.; Nonogaki, H.; Fait, A. Activation and regulation of primary metabolism during seed germination. *Seed Sci. Res.* **2014**, *24*, 1–15. [CrossRef]
24. Mittler, R.; Vanderauwera, S.; Gollery, M.; Van Breusegem, F. Reactive oxygen gene network of plants. *Trends Plant Sci.* **2004**, *9*, 490–498. [CrossRef]
25. Mittler, R. ROS Are Good. *Trends Plant Sci.* **2017**, *22*, 11–19. [CrossRef] [PubMed]
26. Foyer, C.H.; Ruban, A.V.; Noctor, G. Viewing oxidative stress through the lens of oxidative signalling rather than damage. *Biochem. J.* **2017**, *474*, 877–883. [CrossRef] [PubMed]
27. Diaz-Vivancos, P.; Barba-Espín, G.; Hernández, J.A. Elucidating hormonal/ROS networks during seed germination: Insights and perspectives. *Plant Cell Rep.* **2013**, *32*, 1491–1502. [CrossRef]
28. Gomes, M.P.; Garcia, Q.S. Reactive oxygen species and seed germination. *Biologia* **2013**, *68*, 351–357. [CrossRef]
29. Ishibashi, Y.; Kasa, S.; Sakamoto, M.; Aoki, N.; Kai, K.; Yuasa, T.; Hanada, A.; Yamaguchi, S.; Iwaya-Inoue, M. A Role for Reactive Oxygen Species Produced by NADPH Oxidases in the Embryo and Aleurone Cells in Barley Seed Germination. *PLoS ONE* **2015**, *10*, e0143173. [CrossRef]

30. Wojtyla, Ł.; Lechowska, K.; Kubala, S.; Garnczarska, M. Different modes of hydrogen peroxide action during seed germination. *Front. Plant Sci.* **2016**, *7*, 66. [CrossRef]
31. Bailly, C. The signalling role of ROS in the regulation of seed germination and dormancy. *Biochem. J.* **2019**, *476*, 3019–3032. [CrossRef]
32. Diaz-Vivancos, P.; Faize, M.; Barba-Espin, G.; Faize, L.; Petri, C.; Hernández, J.A.; Burgos, L. Ectopic expression of cytosolic superoxide dismutase and ascorbate peroxidase leads to salt stress tolerance in transgenic plums. *Plant Biotechnol. J.* **2013**, *11*, 976–985. [CrossRef]
33. Bradford, M.M. A rapid and sensitive method for the quantitation of microgram quantities of protein utilizing the principle of protein-dye binding. *Anal. Biochem.* **1976**, *72*, 248–254. [CrossRef]
34. Bueso, E.; Serrano, R.; Pallás, V.; Sánchez-Navarro, J.A. Seed tolerance to deterioration in arabidopsis is affected by virus infection. *Plant Physiol. Biochem.* **2017**, *116*, 1–8. [CrossRef]
35. Hilhorst, H.W.M.; Smitt, A.I.; Karssen, C.M. Gibberellin-biosynthesis and -sensitivity mediated stimulation of seed germination of Sisymbrium officinale by red light and nitrate. *Physiol. Plant.* **1986**, *67*, 285–290. [CrossRef]
36. Footitt, S.; Huang, Z.; Clay, H.A.; Mead, A.; Finch-Savage, W.E. Temperature, light and nitrate sensing coordinate Arabidopsis seed dormancy cycling, resulting in winter and summer annual phenotypes. *Plant J.* **2013**, *74*, 1003–1015. [CrossRef] [PubMed]
37. Manz, B.; Müller, K.; Kucera, B.; Volke, F.; Leubner-Metzger, G. Water uptake and distribution in germinating tobacco seeds investigated in vivo by nuclear magnetic resonance imaging. *Plant Physiol.* **2005**, *138*, 1538–1551. [CrossRef]
38. Acosta-Motos, J.R.; Noguera-Vera, L.; Barba-Espín, G.; Piqueras, A.; Hernández, J.A. Antioxidant metabolism and chlorophyll fluorescence during the acclimatisation to Ex vitro conditions of micropropagated Stevia rebaudiana bertoni plants. *Antioxidants* **2019**, *8*, 615. [CrossRef]
39. Díaz-Vivancos, P.; Clemente-Moreno, M.J.; Rubio, M.; Olmos, E.; García, J.A.; Martínez-Gómez, P.; Hernández, J.A. Alteration in the chloroplastic metabolism leads to ROS accumulation in pea plants in response to plum pox virus. *J. Exp. Bot.* **2008**, *59*, 2147–2160. [CrossRef] [PubMed]
40. Díaz-Vivancos, P.; Barba-Espín, G.; Clemente-Moreno, M.J.; Hernández, J.A. Characterization of the antioxidant system during the vegetative development of pea plants. *Biol. Plant.* **2010**, *54*, 76–82. [CrossRef]
41. Kvaratskhelia, M.; Winkel, C.; Naldrett, M.T.; Thorneley, R.N.F. A novel high activity cationic ascorbate peroxidase from tea (Camellia sinensis)—A class III peroxidase with unusual substrate specificity. *J. Plant Physiol.* **1999**, *154*, 273–282. [CrossRef]
42. Yamaguchi, S.; Kamiya, Y.; Sun, T.P. Distinct cell-specific expression patterns of early and late gibberellin biosynthetic genes during Arabidopsis seed germination. *Plant J.* **2001**, *28*, 443–453. [CrossRef] [PubMed]
43. Barba-Espín, G.; Hernández, J.A.; Diaz-Vivancos, P. Role of H_2O_2 in pea seed germination. *Plant Signal Behav.* **2012**, *7*, 193–195. [CrossRef] [PubMed]
44. Finch-Savage, W.E.; Leubner-Metzger, G. Seed dormancy and the control of germination. *New Phytol.* **2006**, *171*, 501–523. [CrossRef]
45. Bailly, C.; El-Maarouf-Bouteau, H.; Corbineau, F. From intracellular signaling networks to cell death: The dual role of reactive oxygen species in seed physiology. *Comptes Rendus Biol.* **2008**, *331*, 806–814. [CrossRef]

seeds

MDPI

Communication

Hydrogen Peroxide Imbibition Following Cold Stratification Promotes Seed Germination Rate and Uniformity in Peach cv. GF305

Gregorio Barba-Espín [1], José A. Hernández [1], Cristina Martínez-Andújar [2] and Pedro Díaz-Vivancos [1,*]

[1] Group of Fruit Trees Biotechnology, Department of Plant Breeding, CEBAS-CSIC, Campus Universitario de Espinardo, 25, 30100 Murcia, Spain; gbespin@cebas.csic.es (G.B.-E.); jahernan@cebas.csic.es (J.A.H.)
[2] Group of Plant Hormones, Department Plant Nutrition, CEBAS-CSIC, Campus Universitario de Espinardo, 25, 30100 Murcia, Spain; cmandujar@cebas.csic.es
* Correspondence: pdv@cebas.csic.es

Abstract: (1) Background: Peach cv. GF305 is commonly used in breeding programs due to its susceptibility to numerous viruses. In this study, we aimed to achieve a methodology for rapid and uniform seed germination of peach cv. GF305 in order to obtain vigorous seedlings; (2) Methods: A combination of cold stratification and H_2O_2 imbibition was tested on peach seeds with or without endocarp. In addition, the levels of non-enzymatic antioxidants ascorbate and glutathione as well as the hormone profile in seedling roots and shoots were determined; (3) Results: We found that H_2O_2 imbibition of peach seeds without endocarp after 8 weeks of stratification increased germination rate and resulted in seedlings displaying good vegetative growth. The H_2O_2 imbibition also affected the levels of ascorbate, glutathione, and the phytohormones abscisic acid and jasmonic acid in peach seedlings; (4) Conclusions: Although stratification periods of 12 weeks have been previously established as being appropriate for this cultivar, we have been able to reduce this stratification time by up to 4 weeks, which may have practical implication in peach nurseries.

Keywords: ascorbate; endocarp removal; hydrogen peroxide; glutathione; peach; phytohormones; seed germination; stratification

Citation: Barba-Espín, G.; Hernández, J.A.; Martínez-Andújar, C.; Díaz-Vivancos, P. Hydrogen Peroxide Imbibition Following Cold Stratification Promotes Seed Germination Rate and Uniformity in Peach cv. GF305. *Seeds* **2022**, *1*, 28–35. https://doi.org/10.3390/seeds1010004

Academic Editor: Valeria Cavallaro

Received: 14 December 2021
Accepted: 1 January 2022
Published: 5 January 2022

1. Introduction

In stone fruit species, stratification (moist chilling of seeds) has been described as the most widely method to break seed dormancy and promote germination. Stratification simulates winter conditions keeping seeds chilled for 3 to 4 months [1]. In order to reduce this waiting period, the application of chemicals and the mechanical removal of the seed coat have been widely used in breeding programs [2]. The germination process is associated with many molecular, metabolic, and cellular events enabling radicle emergence and seedling establishment [3,4]. In both dormant and non-dormant seeds, the crucial role of phytohormones regulating seed dormancy breaking and germination has been long established, with reactive oxygen species (ROS) and hence the antioxidative metabolism closely linked [4]. ROS control many different processes in plants via redox-sensitive proteins that act as sensors and messengers of different regulatory pathways [5]. Seed germination must be included among these processes, with the antioxidative metabolism playing a key role [4,6]. However, the biochemical basis of seed dormancy regulation is still poorly understood [7].

Hydrogen peroxide (H_2O_2) has been described as an enhancer of seed germination in many species [3,4]. Different mechanisms have been suggested to explain the H_2O_2 stimulation of seed germination, with the following being the most common: the production of O_2 for mitochondrial metabolism and respiration as a consequence of H_2O_2 scavenging [8], the facilitation of seed cracking, the oxidation of germination inhibitors [9], and the activation of redox-sensitive proteins, inducing changes at proteome, transcriptome, and hormonal levels [4,10]. In this sense, the decrease of abscisic acid (ABA) levels or its transport impairment from cotyledons to the embryo, as well as the mobilization of seed storage proteins, have been suggested as possible mechanisms underlying seed germination promotion through H_2O_2 [10]. In stone fruit seeds, ABA is the main hormone involved in seed dormancy, and a significant decrease in ABA has been recorded as the stratification time increases [1,11]. Moreover, in pea seeds, a role of H_2O_2 in orchestrating the interplay among phytohormones and the cellular redox state leading to seed germination and seedling establishment has been reported [3,10].

Stimulated germination by exogenous H_2O_2 has been reported on endocarp-less seeds of several *Prunus* species when applied before stratification. In this sense, a significant increase in the percentage and speed of seed germination by H_2O_2 was described in the wild almond species *P. scoparia* and *P. communis* [12] as well as in sweet cherry (*P. avium*) [1]. However, the effect of H_2O_2 on the main non-enzymatic antioxidants glutathione and ascorbate and on the hormone profile in peach (*P. persica*) seedlings has not been previously explored. Achieving a rapid and uniform seed germination and also obtaining vigorous seedlings are key goals for peach breeding programs [2]. In this work, we used the peach cv. GF305, which is commonly used in breeding programs due to its susceptibility to numerous viruses [2]. H_2O_2 imbibition following cold stratification of GF305 was applied in order to increase the germination rate and reduce the stratification time. The levels of ascorbate, glutathione, ABA, 1-aminocyclopropane carboxylic acid (ACC), indol acetic acid (IAA), jasmonic acid (JA), salicylic acid (SA), zeatin-riboside (ZR), and zeatin (Z) were analyzed in the seedlings in order to associate changes in these variables with enhanced germination and seedling growth.

2. Materials and Methods

GF305 seeds were obtained from Pépinières Lafond (Valréas Cedex, France). Seeds (approximately 500) were treated with a 2% tetramethylthiuram disulfide (TMTD) fungicide solution for 30 min and then incubated for 3 days in distilled water at 25 °C in the dark, with the water renewed daily. Then, the seeds were introduced in mesh bags and placed in plastic trays with vermiculite previously moistened in a cold chamber at 5 °C in order to fulfill vernalization requirements. After 4, 6, and 8 weeks of stratification, the endocarp of 50% of the peach seeds was manually removed. Three batches of seeds with endocarp (+ endo) and three without endocarp (− endo) were treated as follows: seeds without imbibition (C); seeds imbibed in distilled water (Im); and seeds imbibed in 10 mM H_2O_2 (ImH_2O_2). For seeds without endocarp, the imbibition lasted for 24 h, whereas for seeds with endocarp, the imbibition lasted for 48 h. Afterwards, the seeds were sowed in 48-cell trays containing peat substrate and incubated in a growth chamber at 25 °C, 70% relative humidity, and 500 μmol m^{-2} s^{-1} white light with a 16/8 h photoperiod (light/dark) for 14 days. Finally, seedlings were divided into shoots and roots and weighed to register the fresh weight (FW). The samples were then snap-frozen in liquid nitrogen and stored at −80 °C for further analyses.

The non-enzymatic antioxidants ascorbate and glutathione were determined as previously described [13–15]. Briefly, samples were homogenized in 1 M $HClO_4$, then centrifuged at 12,000× *g* for 10 min and the pH supernatant was adjusted to 5.5–6 with 5 M K_2CO_3. Then, reduced (GSH) and oxidized (GSSG) glutathione were analyzed using dithio-bis-2-nitrobenzoic acid and glutathione reductase in the presence of NADPH at 412 nm [13–15], whereas reduced ascorbate (ASC) was measured by recording the absorption at 265 nm, and the total ascorbate was determined via oxidation to non-absorbing oxidized ascorbate (DHA) in the presence of ascorbate oxidase [13–15]. Hormones (abscisic acid (ABA), 1-aminocyclopropane carboxylic acid (ACC), indol acetic acid (IAA), zeatin (Z), zeatin-riboside (ZR), salicylic acid (SA), and jasmonic acid (JA)) were extracted from plant tissues and analyzed using a high-performance liquid chromatography/mass spectrometry (HPLC/MS) system consisting of an Agilent 1100 Series HPLC (Agilent Technologies,

Santa Clara, CA, USA) connected to an Agilent Ion Trap XCT Plus mass spectrometer (Agilent Technologies, Santa Clara, CA, USA) following previously published methodology [3]. In order to reduce analysis costs and taking into account that no differences were observed between C and Im seeds in terms of germination rate and seedling development, the determination of non-enzymatic antioxidants and hormones levels were carried out by comparing the seeds submitted to imbibition (Im vs. ImH_2O_2).

The experiments were repeated twice with similar results. Analyses for germination and FW measurements were done on the data of 20–40 specimens, whereas analyses for antioxidants and hormones contents were done on at least three biological replicates, each one based on the pull of shoots or roots of 10 specimens. The data were analyzed by one- or two-way ANOVA using SPSS 22 (IBM Corp., Armonk, NY, USA) software, followed by Duncan's multiple range test ($p \leq 0.05$) in the case of data of germination percentage and seedling FW.

3. Results and Discussion

Cold wet stratification has been widely used for the germination of seeds from *Prunus* species. In this sense, 12 weeks of stratification has been proven to fulfill vernalization requirements in *Prunus*, leading to dormancy breaking and germination percentages near 95% [1,2], whereas decreasing stratification time to 8 weeks reduced germination percentage up to 40% [1]. In this work, we attempted to reduce the stratification time by using H_2O_2 imbibition after stratification. Four weeks of stratification resulted in a very low germination rate (below 10%) with no significant differences among treatments (data not shown). Six weeks of stratification led to germination percentages below 50% in all cases (Figure 1). In this sense, although seeds without endocarp showed significantly higher germination rates than seeds with endocarp, no significant differences among imbibition treatments were found (Figure 1). On the other hand, in 8 weeks-stratified seeds, the imbibition in H_2O_2 remarkably increased the percentage of seeds germination without endocarp up to 86%, compared to non-imbibed seeds (53% germination) and water-imbibed seeds (55% germination; Figure 1). However, the seeds with endocarp showed a lower germination rate (both imbibed and non-imbibed seeds), with the values being statistically comparable to those of 6 weeks-stratified seeds without endocarp (Figure 1). This inhibitory effect of the endocarp on peach seed germination was previously described in peach and could be due to a water uptake delay and the presence of germination inhibitors, such as ABA [2]. According to these results, the subsequent analyses were carried out on seedlings obtained from seeds subjected to 8 weeks of stratification followed by removal of endocarp.

Regarding the seedling growth, in 8 weeks-stratified seeds, the imbibition with H_2O_2 had no effect on it, whereas the imbibition with water after stratification slightly decrease the FW of seedling roots (Figure 2). Thus, after 8 weeks of stratification, H_2O_2-imbibided seeds showed good development and vigor. In comparison to our results, it was previously described that after 12 and 13 weeks of stratification, the resulting plants displayed good development, with no differences between seeds with or without endocarp, whereas a negative effect on seedling growth was observed when a longer period of stratification was applied [2]. Moreover, different authors have pointed out that stratification periods between 10 and 13 weeks were appropriate for peach cultivars [2]. In seeds from wild almond species, the combined treatment of cold stratification with H_2O_2 and GA_3 reduced the time for germination and increased the germination rate, although a synergistic effect was not found [16]. According to these results and our own results, H_2O_2 appears to be an economic and effective agent for large-scale application in seed germination in *Prunus*.

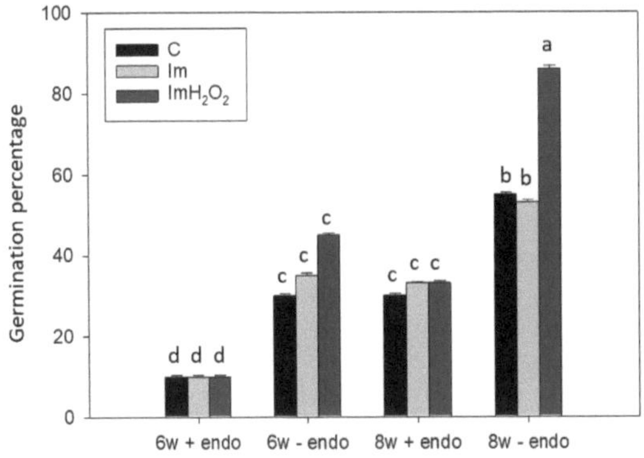

Figure 1. Effect of H_2O_2 imbibition on the germination percentage (%) of peach seeds after 6 (6w) and 8 weeks (8w) of stratification at 5 °C. After stratification, the endocarp of 50% of the peach seeds were manually removed. Three batches of seeds with endocarp (+ endo) or without endocarp (− endo) were treated as follows: seeds without imbibition (C), placed directly from the stratification to the growing trays; seeds imbibed in water (Im); and seeds imbibed in 10 mM H_2O_2 (ImH_2O_2). Seeds were then sowed in trays and incubated at 25 °C with a 16/8-h photoperiod (light/dark) for 14 days. Different letters indicate statistical significance among treatments according to Duncan's test ($p \leq 0.05$).

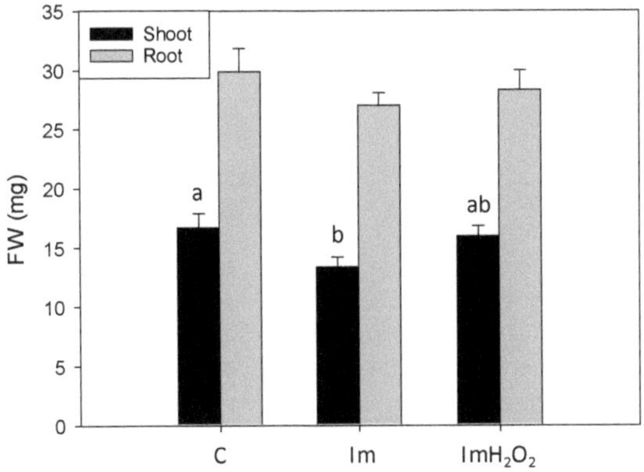

Figure 2. Effect of H_2O_2 treatment on the growth (measured as fresh weight, FW) of peach seedlings resulting from seeds subjected to 8 weeks of stratification followed by endocarp removal. Data represent the mean ± *SE* of at least 20 repetitions. Different letters indicate statistical significance among treatments according to Duncan's test ($p \leq 0.05$).

Seed dormancy is an evolutionary adaptation present in seeds of all temperate fruit species, including peach, that allows seed germination in a favorable season adequate for seedling growth [2]. The presence of a seed coat in stone fruits seeds negatively affects germination, as it constitutes a physical barrier and also contains high levels of ABA [1,2,17]. On the other hand, it has been previously described that H_2O_2 imbibition stimulates seed germination in both dormant and non-dormant seeds, in a manner dependent on the species, as well as the concentration and the timing of application [1–4,6,10,18]. In *P. scoparia*, the combination of cold stratification and 0.5% H_2O_2 was more effective at breaking dormancy than the widely used phytohormone gibberellic acid [12]. This stimulation has been often associated with changes in antioxidative metabolism. In this sense, we observed that in shoots of peach seedlings, H_2O_2 imbibition resulted in a decrease in reduced glutathione (GSH) content, although the glutathione redox state was not affected because the oxidized form (GSSG) also showed a slight decrease (Table 1). In pea seeds, enhanced seedling growth by 20 mM H_2O_2 and 0.25 mM KNO_3 treatments was also correlated with decreased GSH and GSSG levels [3,19]. However, in seedlings roots, an increase in GSH leading to a higher glutathione redox state was recorded following H_2O_2 imbibition (Table 1). Regarding ascorbate levels, H_2O_2 imbibition produced an increase in both reduced ascorbate (ASC) and oxidized ascorbate (DHA) in seedlings shoots, although the differences were not statistically significant (Table 1). In pea seeds treated with different H_2O_2 concentrations, enhanced seedling vigor was correlated with changes in the levels of enzymatic and non-enzymatic antioxidants [3]. The authors observed that H_2O_2 imbibition led to a slight decline in the glutathione and ascorbate redox state due to a GSH decrease and a DHA increase, respectively. Moreover, a rise in ascorbate peroxidase (APX) activity was also recorded in pea seedlings [3]. Similarly, in peach seedling shoots, we observed a decrease in GSH as well as an increase in DHA (Table 1). In this sense, it has been suggested that ascorbate plays a crucial role during seed germination via stimulation of ascorbate biosynthesis and APX activity, although the possibility that they are the consequence rather than the cause of seed vigor cannot be ruled out [20].

Table 1. Effect of water and H_2O_2 imbibition on the ascorbate and glutathione concentrations in the shoots and roots of peach seedlings resulting from seeds submitted to 8 weeks of stratification followed by endocarp removal. The table displays data for reduced and oxidized glutathione (GSH and GSSG, respectively), glutathione redox state (GSH/(GSH+GSSG)), and reduced and oxidized ascorbate (ASC and DHA, respectively).

	GSH nmol^{-1} FW	GSS Gnmol^{-1} FW	GSH/ (GSH+GSSG)	ASC nmol^{-1} FW	DHA nmol^{-1} FW
IM_SHOOT	274.2 ± 13.7	13.1 ± 0.9	0.95 ± 0.01	884.4 ± 62.1	116.6 ± 26.1
IMH_2O_2_SHOOT	211.5 ± 6.9 *	12.2 ± 0.4	0.94 ± 0.00	1184.6 ± 67.0	162.8 ± 39.0
IM_ROOT	105.7 ± 7.4	12.1 ± 0.5	0.89 ± 0.01	nd	nd
IMH_2O_2_ROOT	148.6 ± 4.9 *	12.2 ± 0.4	0.92 ± 0.00 *	nd	nd

Data represent the mean ± *SE* of at least three repetitions. The "*" symbol indicates statistical significance between treatments for either shoots or roots ($p \leq 0.05$). nd: non-detected.

The germination process is linked to important changes in the redox state of the seeds, and a relationship between ROS and plant hormones in this process is well known [21]. It has been widely described that ROS interact in a complex manner with phytohormone networks, triggering signaling pathways that regulate many physiological processes in plants, including seed germination and seedling establishment [4]. In this work, we analyzed the ABA, ACC, IAA, SA, and JA levels and the ratio Z/ZR in shoots and roots of peach seedlings resulting from seeds submitted to 8 weeks of stratification and manual endocarp removal. Seed imbibition with H_2O_2 produced a decrease in ABA and JA in seedling roots. Regarding the rest of the phytohormones, no significant differences were recorded following the H_2O_2 imbibition (Figure 3).

In pea seedlings, H_2O_2 treatment decreased the ABA, IAA, ZR, SA, and JA levels [3]. A decrease in ABA has been traditionally associated with successful seed germination [4,7], with H_2O_2 treatments resulting in a drop in ABA levels [4,10,22], similarly to the one observed in the peach seedling roots (Figure 3). Regarding JA, opposite results have been reported, with either JA inhibiting or promoting the germination process; therefore, the role of JA acid in seed germination is far from being totally understood [4]. Recently, it has been suggested that JA and ABA act synergistically in most of the biological processes, including seed germination [23,24]. A H_2O_2-mediated decrease in ABA and JA levels, such as the one described in pea seeds and seedlings [3,10] as well as peach seedlings (Figure 3), seems to be necessary for seedling growth. In fact, the inhibitory effect of ABA on seed germination in rice was alleviated by impairing JA biosynthesis, suggesting that ABA stimulates JA biosynthesis to then synergistically inhibit seed germination [25]. In this sense, in pea seeds, imbibition with H_2O_2 and ABA overcame the positive effect on seedling growth achieved by H_2O_2 alone in terms of seedling development, which correlated with a decline in the endogenous H_2O_2 level [26].

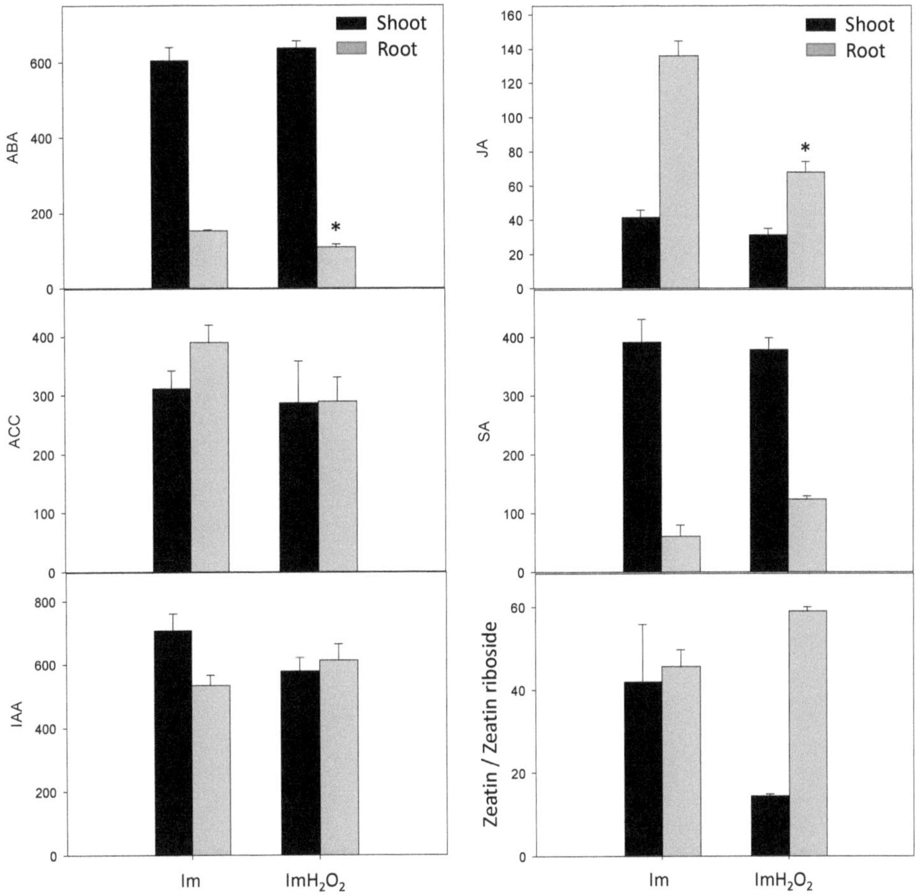

Figure 3. Effect of water and H_2O_2 imbibition on the hormone profile in the shoots and roots of peach seedlings resulting from seeds subjected to 8 weeks of stratification followed by endocarp removal. Data are expressed as nmol g^{-1} FW. Data represent the mean ± *SE* of at least three repetitions. The symbol "*" indicates statistical significance between treatments for either shoots or roots ($p \leq 0.05$).

It has been suggested that keeping the Z/ZR ratio towards the active form (Z) could be important for seedling establishment [3], in a process in which ROS are likely involved in the homeostatic regulation of Z and ZR levels [27]. In this study, the Z/ZR ratio increased in root samples and decreased in shoot samples upon H_2O_2 treatment, although significant differences were not found (Figure 3). In addition to its role in the induction of pathogenesis-related proteins and systemic acquired resistance, the role of SA as a developmental regulator is well reported [28]; however, no significant differences among treatments were found under our experimental conditions. In spite of the well-reported role of ethylene in seed germination and seedling development [4,29], no significant differences were observed in the ethylene precursor ACC (Figure 3), as has also been observed in pea seedlings [3]. However, in soybean, it has been suggested that ROS-induced ethylene production during germination stimulates cell elongation in the root tip [29].

4. Conclusions

In this paper, we have described a method for an efficient and unexpensive reduction of the stratification time required for the germination of peach cv. GF305. After a cold stratification period of 8 weeks, endocarp was removed and seeds were imbibed in 10 mM H_2O_2, resulting in seedlings that displayed good development. Compared to non-treated seeds, for which a stratification period of 12 weeks has been established, we reduced the stratification time by 4 weeks. Moreover, stimulation of seedling growth was also achieved, which correlated with changes in non-enzymatic antioxidants and ABA and JA contents. In general, our findings may have practical application on peach breeding programs and nurseries, as well as on other *Prunus* species.

Author Contributions: Conceptualization: J.A.H. and P.D.-V.; methodology: J.A.H. and G.B.-E.; validation: G.B.-E., P.D.-V. and J.A.H.; formal analysis: G.B.-E. and P.D.-V.; investigation: G.B.-E., P.D.-V. and J.A.H.; resources: J.A.H.; data curation: G.B.-E., P.D.-V. and J.A.H.; writing—original draft preparation: P.D.-V.; writing—review and editing: G.B.-E., P.D.-V., C.M.-A. and J.A.H.; visualization: G.B.-E., P.D.-V. and J.A.H.; supervision: G.B.-E., P.D.-V., C.M.-A. and J.A.H.; project administration: J.A.H. All authors have read and agreed to the published version of the manuscript.

Funding: This research received no external funding.

Institutional Review Board Statement: Not applicable.

Informed Consent Statement: Not applicable.

Data Availability Statement: The data presented in this study are available on request from the corresponding author.

Conflicts of Interest: The authors declare no conflict of interest.

References

1. Stein, M.; Serban, C.; McCord, P. Exogenous ethylene precursors and hydrogen peroxide aid in early seed dormancy release in sweet cherry. *J. Am. Soc. Hortic. Sci.* **2021**, *146*, 50–55. [CrossRef]
2. Martínez-Gómez, P.; Dicenta, F. Mechanisms of dormancy in seeds of peach (Prunus persica (L.) Batsch) cv. GF305. *Sci. Hortic.* **2001**, *91*, 51–58. [CrossRef]
3. Barba-Espin, G.; Diaz-Vivancos, P.; Clemente-Moreno, M.J.; Albacete, A.; Faize, L.; Faize, M.; Pérez-Alfocea, F.; Hernández, J.A. Interaction between hydrogen peroxide and plant hormones during germination and the early growth of pea seedlings. *Plant Cell Environ.* **2010**, *33*, 981–994. [CrossRef] [PubMed]
4. Diaz-Vivancos, P.; Barba-Espin, G.; Hernandez, J.A. Elucidating hormonal/ROS networks during seed germination: Insights and perspectives. *Plant Cell Rep.* **2013**, *32*, 1491–1502. [CrossRef] [PubMed]
5. Foyer, C.H.; Ruban, A.V.; Noctor, G. Viewing oxidative stress through the lens of oxidative signalling rather than damage. *Biochem. J.* **2017**, *474*, 877–883. [CrossRef]
6. Bailly, C.; El-Maarouf-Bouteau, H.; Corbineau, F. From intracellular signaling networks to cell death: The dual role of reactive oxygen species in seed physiology. *Comptes Rendus Biol.* **2008**, *331*, 806–814. [CrossRef] [PubMed]
7. Finch-Savage, W.E.; Leubner-Metzger, G. Seed dormancy and the control of germination. *New Phytol.* **2006**, *171*. [CrossRef] [PubMed]
8. Katzman, L.S.; Taylor, A.G.; Langhans, R.W. Seed enhancements to improve spinach germination. *HortScience* **2001**, *36*, 501–523. [CrossRef]
9. Ogawa, K.; Iwabuchi, M. A mechanism for promoting the germination of Zinnia elegans seeds by hydrogen peroxide. *Plant Cell Physiol.* **2001**, *42*, 286–291. [CrossRef]
10. Barba-Espín, G.; Diaz-Vivancos, P.; Job, D.; Belghazi, M.; Job, C.; Hernández, J.A. Understanding the role of H_2O_2 during pea seed germination: A combined proteomic and hormone profiling approach. *Plant Cell Environ.* **2011**, *34*, 1907–1919. [CrossRef]
11. Leida, C.; Conejero, A.; Arbona, V.; Gómez-Cadenas, A.; Llácer, G.; Badenes, M.L.; Ríos, G. Chilling-dependent release of seed and bud dormancy in peach associates to common changes in gene expression. *PLoS ONE* **2012**, *7*. [CrossRef]
12. Imani, A.; Rasouli, M.; Tavakoli, R.; Zarifi, E.; Fatahi, R.; Barba-ESPÍN, G.; Martínez-Gómez, P. Optimization of seed germination in Prunus species combining hydrogen peroxide or gibberellic acid pre-treatment with stratification. *Seed Sci. Technol.* **2011**, *39*, 204–207. [CrossRef]
13. Diaz-Vivancos, P.; Bernal-Vicente, A.; Cantabella, D.; Petri, C.; Hernández, J.A. Metabolomics and biochemical approaches link salicylic acid biosynthesis to cyanogenesis in peach plants. *Plant Cell Physiol.* **2017**, *58*, 2057–2066. [CrossRef] [PubMed]

14. Pellny, T.K.; Locato, V.; Vivancos, P.D.; Markovic, J.; De Gara, L.; Pallardo, F.V.; Foyer, C.H. Pyridine nucleotide cycling and control of intracellular redox state in relation to poly (ADP-Ribose) polymerase activity and uuclear localization of glutathione during exponential growth of Arabidopsis cells in culture. *Mol. Plant* **2009**, *2*, 442–456. [CrossRef]
15. Vivancos, P.D.; Dong, Y.P.; Ziegler, K.; Markovic, J.; Pallardo, F.V.; Pellny, T.K.; Verrier, P.J.; Foyer, C.H. Recruitment of glutathione into the nucleus during cell proliferation adjusts whole-cell redox homeostasis in Arabidopsis thaliana and lowers the oxidative defence shield. *Plant J.* **2010**, *64*, 825–838. [CrossRef] [PubMed]
16. Zeinalabedini, M.; Majourhat, K.; Khayam-Nekoui, M.; Hernández, J.A.; Martínez-Gómez, P. Breaking seed dormancy in long-term stored seeds from Iranian wild almond species. *Seed Sci. Technol.* **2009**, *37*, 267–275. [CrossRef]
17. Kim, D.H. Practical methods for rapid seed germination from seed coat-imposed dormancy of *Prunus yedoensis*. *Sci. Hortic.* **2019**, *243*, 451–456. [CrossRef]
18. El-Maarouf-Bouteau, H.; Bailly, C. Oxidative signaling in seed germination and dormancy. *Plant Signal. Behav.* **2008**, *3*, 175–182. [CrossRef] [PubMed]
19. Hernández, J.A.; Díaz-Vivancos, P.; Acosta-Motos, J.R.; Barba-Espín, G. Potassium nitrate treatment is associated with modulation of seed water uptake, Antioxidative Metabolism and Phytohormone Levels of Pea Seedlings. *Seeds* **2022**, *1*, 5–15. [CrossRef]
20. De Tullio, M.C.; Arrigoni, O. The ascorbic acid system in seeds: To protect and to serve. *Seed Sci. Res.* **2003**, *13*, 249–260. [CrossRef]
21. Hernández Cortés, J.A. Seed Science Research: Global trends in seed biology and technology. *Seeds* **2022**, *1*, 1–4. [CrossRef]
22. Wang, M.; Van Der Meulen, R.M.; Visser, K.; Van Schalk, H.P.; Van Duijn, B.; De Boer, A.H. Effects of dormancy-breaking chemicals on ABA levels in barley grain embryos. *Seed Sci. Res.* **1998**, *8*, 129–137. [CrossRef]
23. Liu, Z.; Zhang, S.; Sun, N.; Liu, H.; Zhao, Y.; Liang, Y.; Zhang, L.; Han, Y. Functional diversity of jasmonates in rice. *Rice* **2015**, *8*, 42. [CrossRef] [PubMed]
24. Tang, G.; Ma, J.; Hause, B.; Nick, P.; Riemann, M. Jasmonate is required for the response to osmotic stress in rice. *Environ. Exp. Bot.* **2020**, *175*, 104047. [CrossRef]
25. Wang, Y.; Hou, Y.; Qiu, J.; Wang, H.; Wang, S.; Tang, L.; Tong, X.; Zhang, J. Abscisic acid promotes jasmonic acid biosynthesis via a 'SAPK10-bZIP72-AOC' pathway to synergistically inhibit seed germination in rice (Oryza sativa). *New Phytol.* **2020**, *228*, 1336–1353. [CrossRef] [PubMed]
26. Barba-Espin, G.; Nicolas, E.; Almansa, M.S.; Cantero-Navarro, E.; Albacete, A.; Hernandez, J.A.; Diaz-Vivancos, P. Role of thioproline on seed germination: Interaction ROS-ABA and effects on antioxidative metabolism. *Plant Physiol. Biochem.* **2012**, *59*, 30–36. [CrossRef]
27. Gidrol, X.; Lin, W.S.; Dégousée, N.; Yip, S.F.; Kush, A. Accumulation of reactive oxygen species and oxidation of cytokinin in germinating soybean seeds. *Eur. J. Biochem.* **1994**, *224*, 21–28. [CrossRef]
28. Hernández, J.A.; Diaz-Vivancos, P.; Barba-Espín, G.; Clemente-Moreno, M.J. *On the Role of Salicylic Acid in Plant Responses to Environmental Stresses*; Springer: Berlin, Germany, 2017; ISBN 9789811060687.
29. Ishibashi, Y.; Koda, Y.; Zheng, S.H.; Yuasa, T.; Iwaya-Inoue, M. Regulation of soybean seed germination through ethylene production in response to reactive oxygen species. *Ann. Bot.* **2013**, *111*, 95–102. [CrossRef]

seeds

MDPI

Article

Seed Priming Enhances Seed Germination and Morphological Traits of *Lactuca sativa* L. under Salt Stress

Bikash Adhikari, Omolayo J. Olorunwa and T. Casey Barickman *

Department of Plant and Soil Sciences, Mississippi State University, North Mississippi Research and Extension Center, Verona, MS 38879, USA; ba917@msstate.edu (B.A.); ojo26@msstate.edu (O.J.O.)
* Correspondence: t.c.barickman@msstate.edu; Tel.: +1-(662)-566-2201

Abstract: Seed germination is the stage in which plants are most sensitive to abiotic stress, including salt stress (SS). SS affects plant growth and performance through ion toxicity, decreasing seed germination percentage and increasing the germination time. Several priming treatments were used to enhance germination under SS. The objectives of this study were (1) to identify priming treatments to shorten the emergence period, (2) to evaluate priming treatments against the SS, and (3) to induce synchronized seed germination. Salt-sensitive 'Burpee Bibb' lettuce seeds were treated with 0.05% potassium nitrate, 3 mM gibberellic acid, and distilled water. All the primed and non-primed seeds were subjected to 100 mM sodium chloride (NaCl) or 0 mM NaCl (control). The seven-day experiment, arranged in a complete randomized block design with four replications, was conducted in a growth chamber maintained with 16/8 h photoperiod (light/dark), 60% relative humidity, and a day/night temperature of 22/18 °C. The result indicated that hydro-primed (HP) seeds were better synchronized under SS. Similarly, fresh mass (FM) and dry mass (DM) of cotyledon, hypocotyl, and radicle were the highest in HP lettuce regardless of SS. Electrolyte leakage was the lowest in the HP lettuce, while other priming methods under SS increased membrane permeability, leading to osmotic stress and tissue damage. Overall, hydro-priming can be a good priming method for synchronizing germination and increasing FM and DM by creating the least osmotic stress and ion toxicity in lettuce under SS.

Keywords: hydro-priming; lettuce; sodium chloride; synchronization; electrolyte leakage

Citation: Adhikari, B.; Olorunwa, O.J.; Barickman, T.C. Seed Priming Enhances Seed Germination and Morphological Traits of *Lactuca sativa* L. under Salt Stress. *Seeds* **2022**, *1*, 74–86. https://doi.org/10.3390/seeds1020007

Academic Editors: José Antonio Hernández Cortés, Gregorio Barba-Espín and Pedro Diaz-Vivancos

Received: 2 March 2022
Accepted: 29 March 2022
Published: 1 April 2022

1. Introduction

Salt stress (SS) is destructive abiotic stress that affects crop production in arid and semiarid areas [1]. For this reason, it poses a severe threat to food security. At least 20% of crop cultivation land has been damaged by salt accumulation worldwide [2]. It is estimated that around 4000 hectares of land are affected daily and 50% of the arable land is affected by the salt in the soil [3,4]. It is speculated that more than 50% of the cultivable land will turn into non-arable land by the mid-21st century [5]. SS, especially under sodium chloride (NaCl), affects plant growth and performance through ion toxicity and osmotic stress in leaves and roots. Several reports on the relation between SS and seed germination state that SS also suppresses seed germination percentage and increases the germination time [6–8]. Most crops are reported to be sensitive to SS even as low as 3 dSm^{-1} electrical conductivity [9]. The suppression is due to the adverse physiological and biochemical changes in germinating seeds exposed to SS [1]. Specifically, Na^+ and Cl^- impair the seed growth and development through osmotic and ion-specific toxic effects [5,10,11]. Na^+, on the one hand, can replace the K^+ cation and promote ion toxicity, while Cl^- controls the vegetative growth [12].

Seeds are supposed to turn into healthy and vigorous plants with proper metabolisms and physiological performances. To achieve that outcome, seeds maintain low moisture content during their physiological maturity [13]. Lettuce seeds with moisture levels less than 5% can be stored for a long period [14]. However, the extended seed storage period can deteriorate the germination and viability of the seed [15]. Seeds stored for an extended period when subjected to SS can result in even poorer germination and seedling growth. Seed germination and seedling growth are critical stages for crop establishment and are considered the stages in which plants are the most sensitive to abiotic stresses [16,17]. The 'two-phase growth response to SS' concept by Lauchi and Grattan [18] stated that the germination rate is directly correlated with the level of salt treatments applied.

Several seed priming treatments, including hydro- (HP), potassium nitrate- (KNO_3), and gibberellic acid-priming (GA_3), were used to enhance germination and performance under SS [19,20]. The purpose of these priming treatments was to shorten the emergence period and protect the seeds from biotic and abiotic factors during the crucial phase of seedling establishment, synchronizing emergence and leading to a uniform stand and improved yields. Priming is a rehydrating approach that uses different biochemical agents to allow regulated seed rehydration, stimulating metabolic processes that are ordinarily active during the early stages of germination, while preventing the seed from progressing to complete germination [21]. Plant species/genotype and physiology, environmental conditions exposed to the seed lot and vigor, and the priming method used all play roles in the success of seed priming [22]. Priming treatments have some beneficial effects, such as germination enhancement, emergence synchronization, early seedling growth, and minimizing the deleterious effect of abiotic and biotic stress. This has been demonstrated in many crops, such as wheat (*Triticum aestivum*) [23], chickpea (*Cicer arietinum*) [24], sunflower (*Helianthus* pp.) [25], and cotton (*Gossypium* spp.). [26]. Additionally, several other reports on maize (*Zea mays*), broccoli (*Brassica oleracea* var. *italica*), cauliflower (*Brassica oleracea* var. *botrytis*), and other leafy vegetables stated that priming reduced the mean germination time (MGT) and increased the mean germination rate (MGR) and synchronization index (Z) [27,28]. MGT is a measure of the time spread of the germination while MGR is the reciprocal of MGT [29,30]. Similarly, Z is a degree of homogeneity of germination over time [31], which indicates the highest synchronized germination when the Z value is the lowest [32]. These parameters are the measures of germination, seed vigor, and response to abiotic and biotic stress, which help to improve understanding of the roles of different priming methods used under different environmental conditions and on different crops [33,34].

Several studies have been conducted to investigate the influence of different priming methods on different C_3 plants. However, there has been very little information on the interactive effect of various priming methods and salt levels on the germination synchronization, germination parameters, or morphological and biochemical traits in lettuce. Hence, the objectives of this study are (1) to distinguish the priming treatments that can effectively shorten the emergence period, (2) to evaluate the responses of priming treatments under SS, and (3) to induce synchronized seed germination in lettuce. Overall, the goal of this study is to examine priming approaches to overcoming the negative effects of SS on lettuce germination.

2. Materials and Methods

2.1. Planting Materials and Priming Treatment

The salt-susceptible *Lactuca sativa* (cv. 'Burpee Bibb') (BB) seeds were purchased from Burpee (Warminster, PA, USA). The susceptibility of BB was determined from a study by Adhikari et al. [35]. Before sowing, 200 lettuce seeds were treated with three different priming treatments: (1) 0.5% KNO_3, (2) 3 mM GA_3, and (3) hydro-priming. These priming agents were prepared using distilled water. A set of non-primed (NP) seeds were also included for comparison with primed seeds. Seeds were immersed in different priming treatments for different numbers of hours (KNO_3 = 2 h 30 min; GA_3 = 12 h 45 min; and hydro-priming = 2 h 30 min), as described by Mahmoudi et al. [36], with few modifications. Soon after the priming, seeds were rinsed four times with distilled water and were left to dry at room temperature until they regained their original moisture content, which took approximately 2 days (maximum).

2.2. Salt Treatment and Germination

All the primed and non-primed seeds were subjected to either 0 mM NaCl (control) or 100 mM NaCl concentration. The NaCl solution was prepared using distilled water. Before sowing, seeds were soaked in 200 mL of either control or 100 mM NaCl for around 2 h and left to dry for 2 more hours to regain their original moisture content at room temperature. The germination was carried out by placing 25 seeds per treatment per replication in a 10 cm Petri dish with a double-layer filter paper in the growth chamber. The 0 mM NaCl treatment was considered as a control. The growth chamber was maintained with 16/8 h photoperiod (light/dark), 330 μmol/m^2/s of photosynthetic photon flux density (PPFD), 50–60% relative humidity, and a day/night temperature of 22/18 °C. Treatments were arranged in a complete randomized block design with four replications each.

2.3. Germination Parameters and Morphological Traits

2.3.1. Germination Parameters

Seed germination was recorded every 24 h for 5 days. Seeds were considered germinated when the radicle protruded through the seed coat and had a length of at least 2 mm. The mean germination time (MGT), mean germination rate (MGR), germination index (GI), and synchronization index (Z) were measured using the following equations:

$$\text{MGT} = \frac{\sum_{i=1}^{k} n_i \times t_i}{\sum_{i=1}^{k} n_i} \quad [37]$$

where

$n_i t_i$ = seeds germinated at ith interval with the corresponding time interval
n_i = number of seeds germinated in the ith time
ti = time taken for seeds to germinate at ith count

$$\text{MGR} = \frac{1}{MGT} \quad [29]$$

$$\text{GI} = \sum_{i=1}^{k} \frac{n_i}{t_i} \quad [38]$$

$$Z = \frac{\sum_{i=1}^{k} C_{ni,2}}{C_{\sum ni,2}} \quad [31]$$

where

$$Cn_{i,2} = n_i\,(n_i - 1)/2$$

2.3.2. Seeds' Morphological Traits

Five days after sowing, seeds were divided into cotyledon, hypocotyl, and radicle for the determination of growth parameters. Fresh mass (FM), dry mass (DM), and length of all parts of the seeds were recorded.

2.4. Electrolyte Leakage

Electrolyte leakage (EL) was determined using the method described by Quartacci et al. [39]. The cotyledon and the hypocotyl parts of the lettuce plant were cut into 3–4 mm pieces and placed in a 50 mL tube containing 23–30 mL of double-distilled water. The initial electrical conductivity (EC_1) was recorded after two hours using the digital EC meter (Fisherbrand^{TM} accumet AP85 portable waterproof pH/Conductivity meter, Thermo Fisher Scientific, Waltham, MA, USA). The samples were then immersed in liquid nitrogen for 2–3 min and placed back in the same 50 mL tube for an additional 2 h, with continuous shaking. The final electrical conductivity (EC_2) was then recorded. EL was calculated using the following equation:

$$EL\ (\%) = \frac{EC_1}{EC_2}$$

2.5. Data Analysis

Statistical analysis of the data was performed using SAS (version 9.4; SAS Institute, Cary, NC, USA). Data were analyzed using PROC GLM analysis of variance (ANOVA), followed by mean separation. The standard errors were based on the pooled error term from the ANOVA table. Tukey's test ($p \leq 0.05$) was used to differentiate between genotype classifications and treatment.

3. Results

The effects of different priming methods on lettuce seeds subjected to salt stress (SS) were studied through various germination indicators, such as mean germination time (MGT), mean germination rate (MGR), germination index (GI), and synchronization index (Z). These parameters help to achieve a better understanding of seed germination and growth in primed lettuce seeds subjected to SS. Additionally, the electrolyte leakage in germinated seed parts and the morphological traits of cotyledon, hypocotyl, and radicle, including fresh mass (FM), dry mass (DM), and length, were recorded.

3.1. Germination Parameters

None of the priming methods affected MGT, MGR, and GI values in lettuce subjected to SS (Table 1). There was no interaction effect observed between different priming and salt treatments, except for Z ($p < 0.05$). The Z value was recorded as the lowest in HP lettuce under both salt treatment levels. The Z values of other priming methods (GA_3 and KNO_3) were recorded highest; however, they were not significantly different when compared with the NP control.

Table 1. Mean responses of various germination parameters of lettuce seeds for evaluation of the effects of different priming methods (GA_3, hydro-priming, KNO_3, and NP) and two salt treatment levels (control and 100 mM NaCl).

Priming	Treatment	MGT [3]	MGR	Z	GI
GA_3	Control	2.16 ± 0.03a [1,2]	0.5 ± 0.01a	0.72 ± 0.05a	11.8 ± 0.14a
	NaCl	2.23 ± 0.11a	0.4 ± 0.02a	0.65 ± 0.11ab	11.5 ± 0.12a
Hydro	Control	2.01 ± 0.02a	0.5 ± 0.03a	0.32 ± 0.08c	12.5 ± 0.01a
	NaCl	2.04 ± 0.01a	0.4 ± 0.08a	0.42 ± 0.03c	12.3 ± 0.02a
KNO_3	Control	2.17 ± 0.04a	0.5 ± 0.01a	0.71 ± 0.09a	11.8 ± 0.16a
	NaCl	2.37 ± 0.18a	0.4 ± 0.03a	0.55 ± 0.13b	11.2 ± 0.18a
Non-priming	Control	2.17 ± 0.02a	0.5 ± 0.01a	0.71 ± 0.03a	11.8 ± 0.08a
	NaCl	2.32 ± 0.11a	0.4 ± 0.02a	0.56 ± 0.18ab	11.2 ± 0.43a
Priming		***	***	***	***
Salt		**	**	***	***
Priming × Salt treatment		NS	NS	**	NS

[1] Values represent the mean ± SE, $n = 25$. [2] Different lowercase letters indicate significant differences in parameters ($p < 0.05$) as assessed by Tukey's test. NS represents nonsignificant $p > 0.05$; **, *** represent significance levels at $p \leq 0.01$ and $p \leq 0.001$, respectively. [3] MGT = mean germination time (day); MGR = mean germination rate (day^{-1}); GI = germination index (unitless); and Z = synchronization index (unitless).

3.2. Morphological Traits

3.2.1. Fresh Mass

The FM data (Figure 1) indicated a significant (priming × salt treatment) interaction effect on FM of all morphological parameters (hypocotyl, cotyledon, and radicle). Figure 1A demonstrates that hydro-priming resulted in higher cotyledon FM compared to other priming methods (GA_3 and KNO_3). Although there was no significant difference in the hypocotyl FM compared to its control, the hypocotyl FM was recorded highest (Figure 1B) compared to other priming methods, irrespective of the salt treatments. There was no significant effect of GA_3- and KNO_3-priming on hypocotyl FM when treated with salt, compared to the control. Additionally, the cotyledon and radicle FMs of GA_3- and KNO_3-

primed lettuce were not significantly different from non-primed (NP) lettuce under both salt treatment levels. Similarly, there was no significant difference in radicle FM of HP salt-treated lettuce compared to the control (Figure 1C). However, the highest radicle FM was observed for HP lettuce.

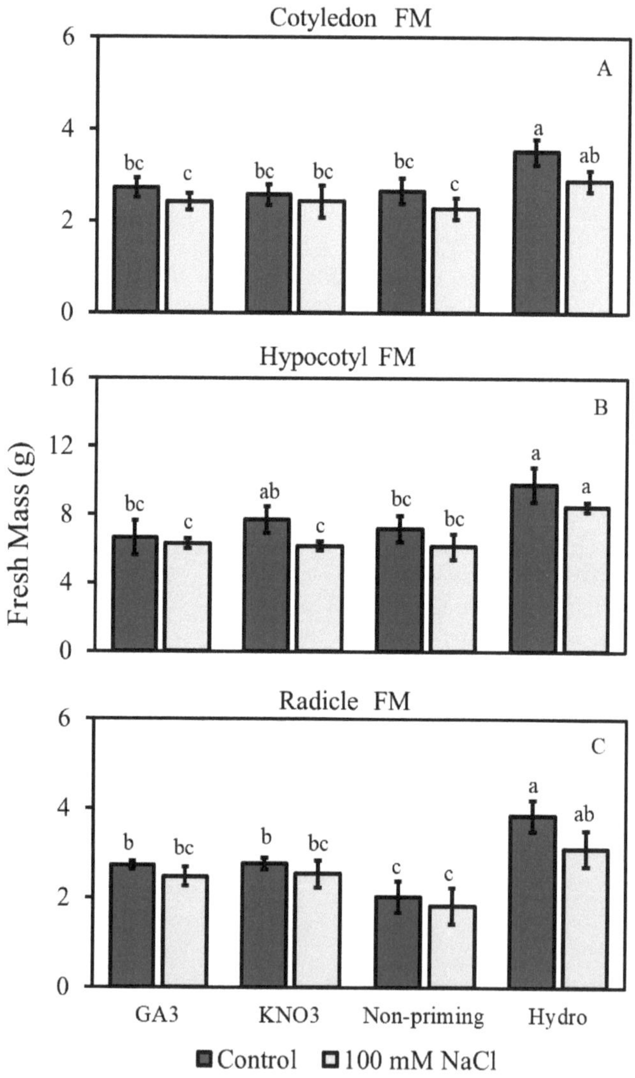

Figure 1. Fresh Mass (FM) of the: (**A**) cotyledon; (**B**) hypocotyl; and (**C**) radicle, (mean ± SE, n = 4), of BB lettuce, non-primed and seed-primed with GA_3, KNO_3, and hydro-priming, and further subjected to 100 mM of NaCl salt stress. 0 mM NaCl was considered as a control. Bars marked with different lowercase letters indicate statistically significant difference using Tukey's honestly difference test at α = 0.05.

3.2.2. Dry Mass

The dry mass (DM) of lettuce responded differently to different priming methods when subjected to two salt treatments, as presented in Figure 2. Priming under different salt treatment levels had a significant interaction effect on cotyledon, hypocotyl, and radicle DMs. Cotyledon DM in HP lettuce was significantly higher compared to other priming methods (GA_3 and KNO_3). Interestingly, there was no significant difference in the cotyledon DM of HP lettuce subjected to SS compared to the control (Figure 2A). It is also worth noting that DM due to hydro-priming increased by almost 100% compared to GA_3-, KNO_3- and NP in both salt treatment levels.

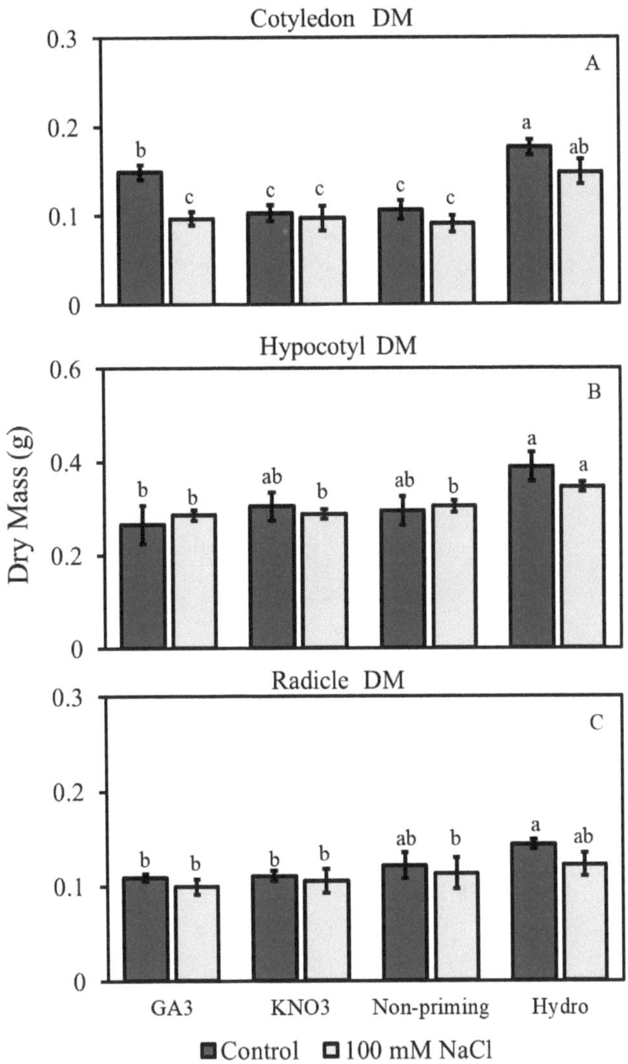

Figure 2. Dry Mass (DM) of the: (**A**) cotyledon; (**B**) hypocotyl; and (**C**) radicle (mean ± SE, $n = 4$) of BB lettuce, non-primed and seed-primed with GA_3, KNO_3, and hydro-priming, and further subjected to 100 mM of NaCl salt stress. 0 mM NaCl was considered as a control. Bars marked with different lowercase letters indicate statistically significant difference using Tukey's honestly difference test at $\alpha = 0.05$.

The DM of hypocotyl was found to be significantly different ($p < 0.05$) in different priming methods studied under different salt treatment levels (Figure 2B). Although no salt treatment effect was observed within the priming methods, a significant interaction effect showed that hypocotyl DM was measured the highest in HP lettuce and the lowest in GA_3-primed lettuce. As in cotyledon DM, priming and salt treatment were significantly affected in the radicle DM. DM of the radicle was recorded highest in the HP lettuce, followed by the NP lettuce (Figure 2C).

3.2.3. Hypocotyl and Radicle Lengths

The priming methods and salt treatment interacted to significantly affect the hypocotyl and radicle lengths (Figure 3). There was a significant decrease ($p < 0.05$) in hypocotyl length by 42%, 49%, and 44% in GA_3-primed, KNO_3-primed, and NP lettuce, respectively, when subjected to SS, as compared to the control (Figure 3A). However, the HP lettuce remained unaffected by the salt treatment compared to the control. The hypocotyl length was recorded highest in the HP lettuce. As in hypocotyl length, a significant decrease in radicle length by 58%, 59%, and 72% was observed in GA_3-primed, KNO_3-primed, and NP lettuce, respectively, under SS as compared to the control (Figure 3B). However, there was no significant decrease (14%) in radicle length recorded in HP and salt-treated lettuce compared to the control. Overall, the length parameters revealed the non-significant suppression in the hypocotyl and radicle length of HP lettuce subjected to 100 mM NaCl.

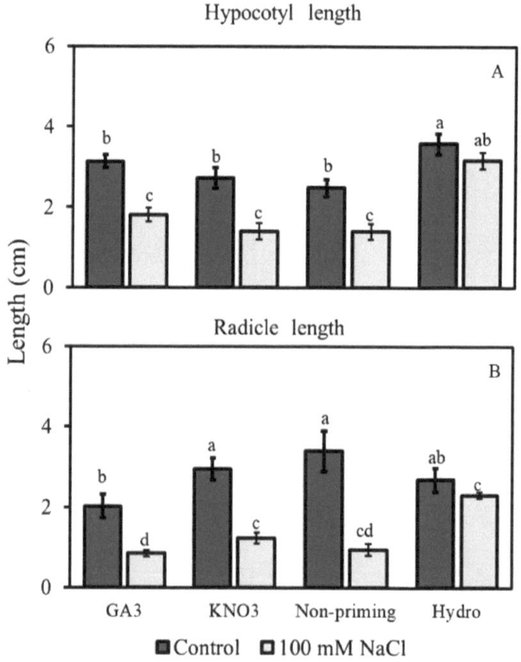

Figure 3. Length of: (**A**) hypocotyl; and (**B**) radicle, (mean ± SE, $n = 4$) of BB lettuce, non-primed and seed-primed with GA_3, KNO_3, and hydro-priming, and further subjected to 100 mM of NaCl salt stress. 0 mM NaCl was considered as a control. Bars marked with different lowercase letters indicate statistically significant difference using Tukey's honestly difference test at $\alpha = 0.05$.

3.3. Electrolyte Leakage

The membrane injury (EL) was noticeably increased in lettuce primed with GA_3, KNO_3, and NP subjected to 100 mM NaCl compared to the control, except for HP lettuce (Figure 4). The increase in EL by 38%, 25%, and 60% was recorded in GA_3, KNO_3, and NP lettuce, respectively, under SS compared to the control. However, there was no significant membrane damage caused in HP lettuce. Salt-treated HP lettuce reflected the lowest EL compared to other salt-treated primed lettuce.

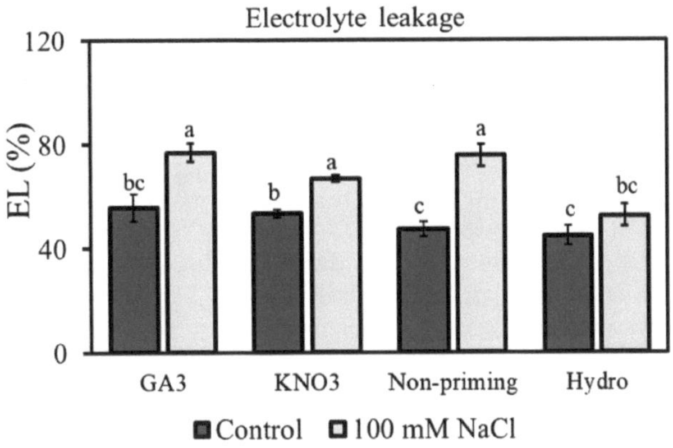

Figure 4. Electrolyte leakage (mean ± SE, n = 4) in BB lettuce, non-primed and seed-primed with GA_3, KNO_3, and hydro-priming, and further subjected to 100 mM of NaCl salt stress. 0 mM NaCl was considered as a control. Bars marked with different lowercase letters indicate statistically significant difference using Tukey's honestly difference test at $\alpha = 0.05$.

4. Discussion

Seed germination and seedling growth are two important stages for crop establishment [17]. These stages are critically sensitive to any abiotic stress, including SS, resulting in poor or delayed germination [7,16]. Lettuce is a moderately salt-tolerant crop (electrical conductivity threshold of 1.3–2.0 dS/m), and can be affected by the osmotic effect of salt in the first phase of salt absorption (osmotic stress phase), followed by the second phase of inhibition of potassium (K^+) ions, the third phase (salt-specific effect phase) of inability to prevent salt ions from accumulating to toxic levels, and the final phase of oxidative stress and cell death [40–42]. During the first phase, the water potential is affected, leading to decreased water uptake by plants, while the second and third phases disturb the ion homeostasis of the cell, leading to membrane disorganization and inhibition of photosynthesis in plants [42]. This four-phase effect of SS was reported in several crops, such as tomato (*Solanum lycopersicum*) [43], wheat (*Triticum aestivum)* [44], sugar beet (*Beta vulgaris*) [45], beans (*Phaseolus vulgaris*) [46], and several other C3 crops [42]. Under normal conditions, the cytosol in plants contains 100 mM of K^+ and less than 10 mM Na^+ [47]. However, under SS, the cytosolic Na^+ and Cl^- can increase up to 100 mM, becoming cytotoxic [47]. A cytotoxic situation leads to protein denaturation, membrane damage, and destabilization, ultimately causing electrolyte leakage and cell death [47]. The current study demonstrated that 100 mM NaCl treatment increases the mean germination time (MGT) and reduces the mean germination rate (MGR). Supporting these results, several reports on SS stated that concentration beyond the crop threshold could decrease the germination rate and increase germination time [6,7,16]. This situation could be caused by the plant's quick response to the high osmotic pressure created by SS, leading to the inhibition of cellular and biosynthetic processes [47]. While the SS caused seed growth inhibition in crops and leafy vegetables with moderate to low salt tolerance, seed priming, in contrast, was reported to improve the overall physiological and morphological performance of those crops [21]. For example, calcium chloride, used on wheat (*Triticum aestivum*) and sorghum (*Sorghum bicolor*); potassium chloride, used on wheat; deionized or distilled water, used on maize (*Zea mays*), barley (*Hordeum vulgare*), and Chinese cabbage (*Brassica rapa* subsp. *Pekinensis*); Gibberellic acid, used on pea (*Pisum sativum*), maize (*Zea mays*), rice (*Oryza sativa*), and alfalfa (*Medicago sativa*); glycine betaine, used on safflower (*Carthamus tinctorius*); potassium nitrate, used on sunflower (*Helianthus* spp.); and potassium chloride, used on broccoli (*Brassica oleracea* var. *italica*), were reported effective in SS amelioration through improved seed germination rate and improved yield [23,48–56]. Priming methods are crop-specific, and a recent study was conducted to demonstrate the difference in the efficacy levels of different priming methods based on several reports in C_3 and C_4 plants.

Previously, it has been reported that synchronized seed growth is also enhanced by the crop-specific seed priming method [11]. In our study, HP seeds had more synchronized seed growth, with the lowest synchronization index (Z), which aligned with the earlier report by Khajeh-Hosseini et al. [11,32]. The current study also demonstrated that hydro-priming inhibited the negative effects of SS on seed germination, while there was a minimal response to KNO_3 and GA_3 of lettuce seeds under SS. Previous research by Mahmoudi et al. [36] found that HP lettuce seedlings could alleviate the NaCl effects on lettuce seedlings better than KNO_3 and GA_3. The NaCl-inhibiting ability of hydro-priming could be due to the ambient or high-water potential (Ψ = 0 MPa) environment created by hydro-priming during the early germination period [47]. Moreover, the decline in synchronization in KNO_3- and GA_3-primed lettuce could be due to a reduction in water potential and cell dehydration caused by SS, leading to a decrease in metabolic activities and protein destabilization during the germination period [47]. Thus, it is worth noting that HP lettuce maintained seed germination and growth in a synchronized pattern even under SS better than the rest of the priming treatments under SS, which agrees with the results reported by Mahmoudi et al. [36].

The salt treatment severely impacted the overall morphological parameters. The use of different priming methods reflected the significantly different growth responses compared to the SS created by 100 mM NaCl. Under SS, FM and DM in KNO_3- and GA_3-treated lettuce were found to be indifferent to those in non-primed (NP) lettuce, and were also reduced significantly, as reported earlier in melon (*Cucumis melo*) [57] and chickpea (*Cicer arietinum*) [58]. The germination and seedling improvement with KNO_3 and GA_3 have previously been documented in 'Vista' lettuce [59] and cabbage (*Brassica oleracea* var. *capitata*) [60]. However, the effectiveness of KNO_3 and GA_3 on salt-stressed BB lettuce was limited in the current study. The previous report documented that poor performance of priming agents can be due to amorphous tissue in the seed coat, which controls the permeability of seeds, disrupting the endogenous osmotic equilibrium [21,61]. The disparity can result in nutritional imbalance and poor seedling growth [62]. Although KNO_3 was reported to be effective in improving the seedling growth and establishment in sunflower (*Helianthus* spp.) and GA_3 was reported to be effective in onion (*Allium cepa*) and sesame (*Sesamum indicum*), the poor performances of KNO_3 and GA_3 suggested that the effects of priming under SS can be crop-specific [63]. On the contrary, the FM and DM of lettuce primed with distilled water (HP) increased significantly, irrespective of the salt treatments. The positive response of hydro-priming even under SS can be due to the stimulatory effect of priming on the mediation of cell division in germinating seeds in the early germination stage [64]. The better performance of HP lettuce in terms of water absorption can also be due to the early commencement of metabolic activity in the seeds before the start of radicle emergence [65]. Additionally, improved water-use efficiency has been documented with hydro-priming [66]. The improved water-use efficiency could be due to the efficient breakdown of stored food reserves in the seed during the second phase (lag phase of seed germination), which initiates the onset of cell wall loosening and expansion for the third phase (post-germination phase) [47]. Hydro-priming was reported to improve seed germination, seedling emergence, and productivity in many field crops [8,24,26,63]. The current study also aligned with a study conducted in romaine lettuce by Mahmoudi et al. [36], which demonstrated that HP seeds exhibited higher adaptive potential under SS compared to KNO_3- and GA_3-primed lettuce seeds. Thus, the use of hydro-priming for lettuce under SS could be an effective priming method to improve the FMs and DMs of different morphological traits in germinated seeds.

Abiotic stress is the major source of reactive oxygen species production (ROS), which deteriorates the growth and performance of plants. Since seed degradation is linked to a loss of structural and metabolic integrity and biochemical aberrations, assessing changes in oxidative stress biomarkers, such as electrolyte leakage (EL), is a good method for figuring out which factors promote seed synchronization [67,68]. EL is prevalent in a range of species, tissues, and cell types, and can be caused by a variety of stresses, including salt stress [69,70]. In the current study, an increase in EL leakage was significantly different in salt-treated lettuce, causing membrane injury, unlike in the control. The most prominent EL was observed in the KNO_3-, GA_3-, and non-primed lettuce subjected to SS. Chiu et al. [71] reported that improvement in germination by any priming might be due to enhanced repair of the membrane. However, KNO_3- and GA_3-priming failed to suppress the ROS induced by SS, leading to significant membrane injury. On the contrary, HP lettuce maintained the suppression of ROS even under SS, thus inhibiting the EL and maintaining the seed germination, as reported by Chiu et al. [71] and Mahmoudi et al. [36]. The outcome of the current study establishes that the positive effect of hydro-priming on EL in lettuce could be related to the higher FM and DM reduction, as well as the lowest synchronization index in the presence of SS, thus sparing few electrons for the production of reactive oxygen species [72]. In this study, the FMs and DMs of seed parts were statistically similar in KNO_3- and GA_3-primed seeds, while HP lettuce was significantly different and highest compared to the NP control. This suggests that BB lettuce can ensure better seed growth and germination with HP seeds even under SS, but not with KNO_3- or GA_3-primed seeds. In addition, the lowest seed synchronization index and electrolyte damage due to hydro-priming provides enough evidence to validate the positive impact of hydro-priming in seed germination and synchronization with minimal cell damage. Overall, this research suggests that hydro-priming has a mitigating impact on membrane injury in plants exposed to SS.

5. Conclusions

This study investigated the effects of different priming methods on salt-sensitive BB lettuce cultivars subjected to SS. Seed priming has the potential to improve seed germination and establishment by initiating germination metabolism without radicle protrusion. Thus, the current report shows the benefits of seed priming in the improvement of productivity in several field crops [73]. In addition, seed priming is cost- and resource-effective [73]. To validate the response of several priming methods in lettuce, this research was conducted to demonstrate that HP seeds showed better synchronization under both salt treatment levels than any other primed seeds. Similarly, the FMs and DMs of cotyledons, hypocotyls, and radicles were recorded the highest in HP lettuce, irrespective of salt treatments. It is important to note that low electrolyte leakage could be attributed to the better seed synchronization and higher FM and DM in HP lettuce under SS. After analyzing the overall morphological, biochemical, and germination traits, it is concluded that HP lettuce performed better under SS compared to lettuce primed with KNO_3 and GA_3. Based on these results, using hydro-priming for lettuce subjected to SS can be a suitable priming method to ameliorate the deleterious effect of SS and enhance the synchronized seed germination. In addition, hydro-priming can also be beneficial in stimulating various signaling cascades during the early growth phase, producing faster and more efficient defense responses in lettuce. Thus, suggesting hydro-priming as an effective priming technique in lettuce tends to benefit a grower in terms of better growth synchronization and higher FM and DM.

Author Contributions: B.A.: conceptualization, methodology, validation, formal analysis, investigation, writing—original draft, writing—review and editing, visualization, methodology, validation, investigation. O.J.O.: methodology, validation, formal analysis, investigation. T.C.B.: conceptualization, methodology, validation, formal analysis, investigation, resources, data curation, writing review and editing, visualization, supervision, project administration, funding acquisition. All authors have read and agreed to the published version of the manuscript.

Funding: The project funding is based on work that is supported by the USDA-NIFA Hatch Project under accession number 149210.

Institutional Review Board Statement: Not applicable.

Informed Consent Statement: Not applicable.

Data Availability Statement: The data presented in this study are available on request from the corresponding author.

Acknowledgments: We thank Thomas Horgan and Skyler Brazel at the Vegetable Physiology Laboratory for their technical assistance for their help during data collection.

Conflicts of Interest: The authors declare that they have no known competing financial interest or personal relationships that could have appeared to influence the work reported in this paper.

References

1. Ibrahim, E.A. Seed priming to alleviate salinity stress in germinating seeds. *J. Plant Physiol.* **2016**, *192*, 38–46. [CrossRef] [PubMed]
2. Hussain, M.; Ahmad, S.; Hussain, S.; Lal, R.; Ul-Allah, S.; Nawaz, A. Rice in saline soils: Physiology, biochemistry, genetics, and management. In *Advances in Agronomy*; Sparks, D.L., Ed.; Academic Press: Cambridge, MA, USA, 2018; Volume 148, pp. 231–287.
3. Zaman, M.; Shahid, S.A.; Pharis, R.P. Salinity a Serious Threat to Food Security—Where Do We Stand? *Int. At. Energy Agency Bull.* **2016**, *39*, 9–10.
4. Park, H.J.; Kim, W.Y.; Yun, D.J. A New Insight of Salt Stress Signaling in Plant. *Mol. Cells* **2016**, *39*, 447–459. [CrossRef] [PubMed]
5. WMO. *State of the Global Climate 2020: Provisional Report*; WMO: Geneva, Switzerland, 2021; ISBN 4146702018922.
6. Rouhi, H.R.; Aboutalebian, M.A.; Sharif-Zadeh, F. Effects of hydro and osmopriming on drought stress tolerance during germination in four grass species. *Int. J. AgriScience* **2011**, *1*, 107–114.
7. Jamil, M.; Lee, C.C.; Rehman, S.U.; Lee, D.B.; Ashraf, M.; Rha, E.S. Salinity (NaCl) tolerance of Brassica species at germination and early seedling growth. *Electron. J. Environ. Agric. Food Chem.* **2005**, *4*, 970–976.
8. Omid, A.; Farzad, S.-Z. Osmo and hydro priming improvement germination characteristics and enzyme activity of Mountain Rye (Secale montanum) seeds under drought stress. *J. Stress Physiol. Biochem.* **2012**, *8*, 253–261.
9. Francois, L.E.; Maas, E.V. Crop response and management of salt-affected soils. In *Handbook of Plant and Crop Stress*; Marcel Dekker Press Inc.: New York, NY, USA, 1999; pp. 169–201.
10. Mahmoudi, H.; Kaddour, R.; Huang, J.; Nasri, N.; Olfa, B.; M'Rah, S.; Hannoufa, A.; Lachaâl, M.; Ouerghi, Z. Varied tolerance to NaCl salinity is related to biochemical changes in two contrasting lettuce genotypes. *Acta Physiol. Plant.* **2011**, *33*, 1613–1622. [CrossRef]
11. Khajeh-Hosseini, M.; Powell, A.A.; Bingham, I.J. The interaction between salinity stress and seed vigour during germination of soyabean seeds. *Seed Sci. Technol.* **2003**, *31*, 715–725. [CrossRef]
12. Jenks, M.A.; Hasegawa, P.M.; Jain, S.M. *Advances in Molecular Breeding toward Drought and Salt Tolerant Crops*; Springer: Dordrecht, The Netherlands, 2007; ISBN 1402055773.
13. Tekrony, D.M.; Egli, D.B. Accumulation of seed vigour during development and maturation. In *Basic and Applied Aspects of Seed Biology*; Springer: Berlin/Heidelberg, Germany, 1997; pp. 369–384.
14. Walters, C.; Towill, L.; Collins, C.O. Seeds and pollen. In *Commercial Storage Fruits, Vegetables, and Florist and Nursery Stocks*; United States Department of Agriculture: Washington, DC, USA, 2004; p. 735.
15. Shaban, M. Review on physiological aspects of seed deterioration. *Int. J. Agric. Crop Sci.* **2013**, *6*, 627–631.
16. Yadav, P.V.; Maya, K.; Zakwan, A. Seed priming mediated germination improvement and tolerance to subsequent exposure to cold and salt stress in capsicum. *Res. J. Seed Sci.* **2011**, *4*, 125–136. [CrossRef]
17. Hubbard, M.; Germida, J.; Vujanovic, V. Fungal endophytes improve wheat seed germination under heat and drought stress. *Botany* **2012**, *90*, 137–149. [CrossRef]
18. Läuchli, A.; Grattan, S.R. Plant growth and development under salinity stress. In *Advances in Molecular Breeding toward Drought and Salt Tolerant Crops*; Springer: Dordrecht, The Netherlands, 2007; pp. 1–32.
19. Taylor, A.G.; Harman, G.E. Concepts and technologies of selected seed treatments. *Annu. Rev. Phytopathol.* **1990**, *28*, 321–339. [CrossRef]
20. Afzal, I.; Basra, S.M.A.; Farooq, M.; Nawaz, A. Alleviation of salinity stress in spring wheat by hormonal priming with ABA, salicylic acid and ascorbic acid. *Int. J. Agric. Biol.* **2006**, *8*, 23–28.
21. Paparella, S.; Araújo, S.S.; Rossi, G.; Wijayasinghe, M.; Carbonera, D.; Balestrazzi, A. Seed priming: State of the art and new perspectives. *Plant Cell Rep.* **2015**, *34*, 1281–1293. [CrossRef]
22. Parera, C.A.; Cantliffe, D.J. Presowing seed priming. *Hortic. Rev.* **1994**, *16*, 109–141.
23. Iqbal, M.; Ashraf, M.; Jamil, A.; Ur-Rehman, S. Does seed priming induce changes in the levels of some endogenous plant hormones in hexaploid wheat plants under salt stress? *J. Integr. Plant Biol.* **2006**, *48*, 181–189. [CrossRef]
24. Kaur, S.; Gupta, A.K.; Kaur, N. Effect of osmo-and hydropriming of chickpea seeds on seedling growth and carbohydrate metabolism under water deficit stress. *Plant Growth Regul.* **2002**, *37*, 17–22. [CrossRef]
25. Kaya, M.D.; Okçu, G.; Atak, M.; Cıkılı, Y.; Kolsarıcı, Ö. Seed treatments to overcome salt and drought stress during germination in sunflower (*Helianthus annuus* L.). *Eur. J. Agron.* **2006**, *24*, 291–295. [CrossRef]
26. Casenave, E.C.; Toselli, M.E. Hydropriming as a pre-treatment for cotton germination under thermal and water stress conditions. *Seed Sci. Technol.* **2007**, *35*, 88–98. [CrossRef]
27. Wu, L.; Huo, W.; Yao, D.; Li, M. Effects of solid matrix priming (SMP) and salt stress on broccoli and cauliflower seed germination and early seedling growth. *Sci. Hortic.* **2019**, *255*, 161–168. [CrossRef]

28. Singh, H.; Jassal, R.K.; Kang, J.S.; Sandhu, S.S.; Kang, H.; Grewal, K. Seed priming techniques in field crops—A review. *Agric. Rev.* **2015**, *36*, 251–264. [CrossRef]
29. Ranal, M.A.; de Santana, D.G.; Ferreira, W.R.; Mendes-Rodrigues, C. Calculating germination measurements and organizing spreadsheets. *Braz. J. Bot.* **2009**, *32*, 849–855. [CrossRef]
30. Soltani, E.; Ghaderi-Far, F.; Baskin, C.C.; Baskin, J.M. Problems with using mean germination time to calculate rate of seed germination. *Aust. J. Bot.* **2015**, *63*, 631–635. [CrossRef]
31. Labouriau, L.G. On the frequency of isothermal germination in seeds of *Dolichos biflorus* L. *Plant Cell Physiol.* **1978**, *19*, 507–512. [CrossRef]
32. Ranal, M.A.; Santana, D.G. How and why to measure the germination process? *Braz. J. Bot.* **2006**, *29*, 1–11. [CrossRef]
33. Mavi, K.; Demir, I.; Matthews, S. Mean germination time estimates the relative emergence of seed lots of three cucurbit crops under stress conditions. *Seed Sci. Technol.* **2010**, *38*, 14–25. [CrossRef]
34. Damalas, C.A.; Koutroubas, S.D.; Fotiadis, S. Hydro-priming effects on seed germination and field performance of faba bean in spring sowing. *Agriculture* **2019**, *9*, 201. [CrossRef]
35. Adhikari, B.; Olorunwa, O.J.; Wilson, J.C.; Barickman, T.C. Morphological and Physiological Response of Different Lettuce Genotypes to Salt Stress. *Stresses* **2021**, *1*, 285–304. [CrossRef]
36. Mahmoudi, H.; Massoud, R.B.; Baatour, O.; Tarchoune, I.; Salah, I.B.; Nasri, N.; Abidi, W.; Kaddour, R.; Hannoufa, A.; Lachaâl, M. Influence of different seed priming methods for improving salt stress tolerance in lettuce plants. *J. Plant Nutr.* **2012**, *35*, 1910–1922. [CrossRef]
37. Ellis, R.H.; Roberts, E.H. The quantification of ageing and survival in orthodox seeds. *Seed Sci. Technol.* **1981**, *9*, 373–409.
38. AOSA (Association of Official Seed Analysts). SCST (Society of Commercial Seed Technologists). Rules for testing seeds. *J. Seed Technol.* **1993**, *16*, 1–113.
39. Quartacci, M.F.; Glišić, O.; Stevanović, B.; Navari-Izzo, F. Plasma membrane lipids in the resurrection plant *Ramonda serbica* following dehydration and rehydration. *J. Exp. Bot.* **2002**, *53*, 2159–2166. [CrossRef] [PubMed]
40. Munns, R.; James, R.A.; Läuchli, A. Approaches to increasing the salt tolerance of wheat and other cereals. *J. Exp. Bot.* **2006**, *57*, 1025–1043. [CrossRef] [PubMed]
41. Jamil, A.; Riaz, S.; Ashraf, M.; Foolad, M.R. Gene expression profiling of plants under salt stress. *Crit. Rev. Plant Sci.* **2011**, *30*, 435–458. [CrossRef]
42. Ahmad, P.; Azooz, M.M.; Prasad, M.N.V. *Salt Stress in Plants*; Springer: Berlin/Heidelberg, Germany, 2013.
43. Szepesi, Á. Role of salicylic acid pre-treatment on the acclimation of tomato plants to salt-and osmotic stress. *Acta Biol. Szeged.* **2005**, *49*, 123–125.
44. James, R.A.; Von Caemmerer, S.; Condon, A.G.T.; Zwart, A.B.; Munns, R. Genetic variation in tolerance to the osmotic stress componentof salinity stress in durum wheat. *Funct. Plant Biol.* **2008**, *35*, 111–123. [CrossRef]
45. Gzik, A. Accumulation of proline and pattern of α-amino acids in sugar beet plants in response to osmotic, water and salt stress. *Environ. Exp. Bot.* **1996**, *36*, 29–38. [CrossRef]
46. Gadallah, M.A.A. Effects of proline and glycinebetaine on Vicia faba responses to salt stress. *Biol. Plant.* **1999**, *42*, 249–257. [CrossRef]
47. Taiz, L.; Zeiger, E.; Møller, I.M.; Murphy, A. *Plant Physiology and Development*; Sinauer Associates Incorporated: Sunderland, MA, USA, 2015; ISBN 1605353531.
48. Alasvandyari, F.; Mahdavi, B.; Hosseini, S.M. Glycine betaine affects the antioxidant system and ion accumulation and reduces salinity-induced damage in safflower seedlings. *Arch. Biol. Sci.* **2017**, *69*, 139–147. [CrossRef]
49. Tabassum, T.; Farooq, M.; Ahmad, R.; Zohaib, A.; Wahid, A. Seed priming and transgenerational drought memory improves tolerance against salt stress in bread wheat. *Plant Physiol. Biochem.* **2017**, *118*, 362–369. [CrossRef]
50. Tsegay, B.A.; Andargie, M. Seed priming with gibberellic acid (GA3) alleviates salinity induced inhibition of germination and seedling growth of *Zea mays* L., *Pisum sativum* var. *abyssinicum* A. Braun and *Lathyrus sativus* L. *J. Crop Sci. Biotechnol.* **2018**, *21*, 261–267. [CrossRef]
51. Imran, M.; Boelt, B.; Mühling, K. Zinc seed priming improves salt resistance in maize. *J. Agron. Crop Sci.* **2018**, *204*, 390–399. [CrossRef]
52. Tahira, T.; Riaz, A.; Muhammad, F.; Basra, S.M.A. Improving salt tolerance in barley by osmopriming and biopriming. *Int. J. Agric. Biol.* **2018**, *20*, 2455–2464.
53. Kim, I.-J.L.; Yun, B.-W.; Jamil, M. GA mediated OsZAT-12 expression improves salt resistance of rice. *Int. J. Agric. Biol* **2016**, *18*, 330–336.
54. Bajehbaj, A.A. The effects of NaCl priming on salt tolerance in sunflower germination and seedling grown under salinity conditions. *Afr. J. Biotechnol.* **2010**, *9*, 1764–1770.
55. Younesi, O.; Moradi, A. Effect of priming of seeds of *Medicago sativa* 'bami'with gibberellic acid on germination, seedlings growth and antioxidant enzymes activity under salinity stress. *J. Hortic. Res.* **2014**, *22*, 167–174. [CrossRef]
56. Hassini, I.; Martinez-Ballesta, M.C.; Boughanmi, N.; Moreno, D.A.; Carvajal, M. Improvement of broccoli sprouts (*Brassica oleracea* L. var. italica) growth and quality by KCl seed priming and methyl jasmonate under salinity stress. *Sci. Hortic.* **2017**, *226*, 141–151. [CrossRef]

57. Farooq, M.; Basra, S.M.A.; Rehman, H.; Ahmad, N.; Saleem, B.A. Osmopriming improves the germination and early seedling growth of melons (*Cucumis melo* L.). *Pak. J. Agric. Sci.* **2007**, *44*, 529–536.
58. Sarwar, N.; Yousaf, S.; Jamil, F.F. Induction of salt tolerance in chickpea by using simple and safe chemicals. *Pak. J. Bot.* **2006**, *38*, 325.
59. Nasri, N.; Kaddour, R.; Mahmoudi, H.; Baatour, O.; Bouraoui, N.; Lachaâl, M. The effect of osmopriming on germination, seedling growth and phosphatase activities of lettuce under saline condition. *Afr. J. Biotechnol.* **2011**, *10*, 14366–14372.
60. Jamil, M.; Ashraf, M.; Rha, E. Alleviation of salt stress using gibberellic acid in Chinese cabbage. *Acta Agron. Hung.* **2012**, *60*, 345–355. [CrossRef]
61. Zhou, J.; Wang, Y.; Jahufer, Z. Location and chemical composition of semi-permeable layer of forage seeds. *Bangladesh J. Bot.* **2013**, *42*, 23–30. [CrossRef]
62. Bradford, K.J. A water relations analysis of seed germination rates. *Plant Physiol.* **1990**, *94*, 840–849. [CrossRef] [PubMed]
63. Sahoo, S.; Borgohain, P.; Saha, B.; Moulick, D.; Tanti, B.; Panda, S.K. Seed priming and seedling pre-treatment induced tolerance to drought and salt stress: Recent advances. In *Priming and Pretreatment of Seeds and Seedlings;* Springer: Singapore, 2019; pp. 253–263.
64. Sivritepe, N.; Sivritepe, H.O.; Eris, A. The effects of NaCl priming on salt tolerance in melon seedlings grown under saline conditions. *Sci. Hortic.* **2003**, *97*, 229–237. [CrossRef]
65. Aschermann-Koch, C.; Hofmann, P.; Steiner, A.M. Presowing treatment for improving seed quality in cereals. Germination and vigour. *Seed Sci. Technol.* **1992**, *20*, 435–440.
66. McDonald, M.B. Seed priming. In *Seed Technology and Its Biological Basis;* Sheffild Academic Press: Sheffild, UK, 2000; pp. 287–325.
67. Adetunji, A.E.; Varghese, B.; Pammenter, N.W. Effects of inorganic salt solutions on vigour, viability, oxidative metabolism and germination enzymes in aged cabbage and lettuce seeds. *Plants* **2020**, *9*, 1164. [CrossRef]
68. Al-Maskri, A.; Khan, M.M.; Al-Manther, O.; Al-Habsi, K. Effect of accelerated aging on lipid peroxidation, leakage and seedling vigor (RGR) in cucumber (*Cucumis sativus* L.) seeds. *Pak. J. Agric. Sci.* **2002**, *39*, 330–337.
69. Shabala, S.; Demidchik, V.; Shabala, L.; Cuin, T.A.; Smith, S.J.; Miller, A.J.; Davies, J.M.; Newman, I.A. Extracellular Ca^{2+} ameliorates NaCl-induced K^+ loss from Arabidopsis root and leaf cells by controlling plasma membrane K^+-permeable channels. *Plant Physiol.* **2006**, *141*, 1653–1665. [CrossRef]
70. Demidchik, V.; Maathuis, F. *Ion Channels and Plant Stress Responses;* Springer: Berlin/Heidelberg, Germany, 2010; ISBN 3642104940.
71. Chiu, K.Y.; Wang, C.S.; Sung, J.M. Lipid peroxidation and peroxide-scavenging enzymes associated with accelerated aging and hydration of watermelon seeds differing in ploidy. *Physiol. Plant.* **1995**, *94*, 441–446. [CrossRef]
72. Khan, M.N.; Zhang, J.; Luo, T.; Liu, J.; Rizwan, M.; Fahad, S.; Xu, Z.; Hu, L. Seed priming with melatonin coping drought stress in rapeseed by regulating reactive oxygen species detoxification: Antioxidant defense system, osmotic adjustment, stomatal traits and chloroplast ultrastructure perseveration. *Ind. Crops Prod.* **2019**, *140*, 111597. [CrossRef]
73. Hasanuzzaman, M.; Fotopoulos, V. *Priming and Pretreatment of Seeds and Seedlings: Implication in Plant Stress Tolerance and Enhancing Productivity in Crop Plants;* Springer: Singapore, 2019; ISBN 9789811386244.

seeds

MDPI

Article

Alleviation of Salt-Inhibited Germination and Seedling Growth of Kidney Bean by Seed Priming and Exogenous Application of Salicylic Acid (SA) and Hydrogen Peroxide (H_2O_2)

Shaila Shermin Tania [1,†], Mohammad Saidur Rhaman [1,*,†], Farjana Rauf [1], Md. Moklasur Rahaman [1], Muhammad Humayun Kabir [1], Md. Anamul Hoque [2] and Yoshiyuki Murata [3]

1 Department of Seed Science and Technology, Bangladesh Agricultural University, Mymensingh 2202, Bangladesh; shailasst04@bau.edu.bd (S.S.T.); farjanaraufrakhi@gmail.com (F.R.); moklasurrahaman345@gmail.com (M.M.R.); humayunhort70@gmail.com (M.H.K.)
2 Department of Soil Science, Bangladesh Agricultural University, Mymensingh 2202, Bangladesh; anamul71@bau.edu.bd
3 Department of Bio-Functional Chemistry, Okayama University, Okayama 700-8530, Japan; muta@cc.okayama-u.ac.jp
* Correspondence: saidursst@bau.edu.bd
† These authors contributed equally to this work.

Citation: Tania, S.S.; Rhaman, M.S.; Rauf, F.; Rahaman, M.M.; Kabir, M.H.; Hoque, M.A.; Murata, Y. Alleviation of Salt-Inhibited Germination and Seedling Growth of Kidney Bean by Seed Priming and Exogenous Application of Salicylic Acid (SA) and Hydrogen Peroxide (H_2O_2). *Seeds* **2022**, *1*, 87–98. https://doi.org/10.3390/seeds1020008

Academic Editors: José Antonio Hernández Cortés, Gregorio Barba-Espín and Pedro Diaz-Vivancos

Received: 24 February 2022
Accepted: 6 April 2022
Published: 8 April 2022

Abstract: Salinity is a dominant obstacle to the proper germination of seeds, growth of seedlings, and, consequently, the production of crops. The priming of seeds with different treating agents can efficiently impart salinity tolerance. Kidney bean is a nutritious and popular vegetable crop in the world. Literature shows that salt stress negatively disturbs the germination and growth of kidney beans. In the present research, we investigated the potentiality of salicylic acid (SA) and hydrogen peroxide (H_2O_2) as priming and exogenous agents to alleviate the salinity-inhibited germination and growth of kidney beans. The seeds were pretreated with SA (1 mM and 2 mM) and H_2O_2 (0.1 mM and 0.15 mM) and soaked in normal tap water (hydro-priming) for 60 min. In addition, for the control experiment, untreated seeds were used. Finally, primed seeds were subjected to salt stress (150 mM NaCl). Our results exhibited that salt stress considerably lowered the percentage of germination (GP), germination index (GI), seed vigor index (SVI), shoot length (SL), root length (RL), shoot–root fresh and dry biomass, and plant growth. The results also exhibited that salt stress significantly decreased the relative water content (RWC) and photosynthetic pigments such as chlorophyll, carotenoids, lycopene, and beta-carotene contents. The SA- and H_2O_2- and hydro-priming stimulated the GP, GI, SL, RL, SVI, and seedling growth. Data also revealed that the supplementation of SA and H_2O_2 enhanced RWC and photosynthetic pigments. When compared to other treatments, pretreatment with 1 mM SA was determined to be comparatively more effective at imparting the salt tolerance of kidney beans. Overall, these results, via a heatmap and principal component analysis, uncovered that priming and exogenous applications of SA and H_2O_2 can improve salt tolerance and enhance germination and seedling characteristics of kidney beans.

Keywords: abiotic stress; germination; priming; salinity; photosynthetic pigments

1. Introduction

Successful seed germination and strong seedling vigor are critical determinants for crop growth in their success, as these parameters lead to consistent plant growth and, consequently, high production [1]. As a result, increasing seed vigor and obtaining proper germination is the main goal of the seed grower in order to improve the vital and yield-defining stage of crop establishment. However, different biotic and abiotic stresses hamper seed germination, vigor, seedling characteristics, and plant growth [2]. Drought, salt, and severe temperatures are examples of abiotic stresses that create osmotic stress in crop plants, resulting in an imbalance at the cellular, molecular, and physiological levels, ultimately leading to plant mortality [3]. Salinity is one of the injurious abiotic stresses that extremely affect the germination, growth, and output of crops [4]. Excessive salt levels in cultivated fields are a big worldwide issue at the moment. According to recent statistics, salt disrupts approximately 0.80 billion hectares of terrestrial regions, severely restraining the usage of land for agricultural purposes [5]. From an agricultural standpoint, salinity levels influence approximately 20% of the crop-cultivable area and approximately 33% of irrigated cropland to variable degrees, and by 2050, this figure will surpass 50% [6]. To face salt stress, plants have evolved diverse mechanisms and plant growers are applying different tactics such as seed priming, the supplementation of plant growth regulators, organic fertilizers, and the screening of the right varieties [7–10].

Priming is a simple technology that hydrates seeds to a point where metabolic activity for germination is initiated, but the emergence of a radicle does not arise [11,12]. This method can help a variety of crops whose germination and emergence are hampered by unfavorable soil conditions. Priming stimulates germination by inducing a variety of metabolic changes in the seed and increases seed vitality, resulting in quick and homogenous emergence as well as strong stand establishment [13]. Moreover, seed priming mitigates abiotic stresses in plants by regulating antioxidant enzyme activities, ionic homeostasis, and photosynthetic attributes [14]. Furthermore, priming leads to a faster and stronger induction of basal resistance mechanisms in response to subsequent pathogen attacks, as well as increased tolerance to abiotic stimuli [15,16]. Different signaling molecules and plant hormones used in priming have been well-reported, as have agents that enhance the germination rate and seedling emergence in crops [17,18]. Salicylic acid (SA) is a phenolic endogenous growth regulator that influences a variety of plant progressions, including germination, growth, photosynthesis, and stress tolerance [19,20]. On the other hand, hydrogen peroxide (H_2O_2) performs a vital role in signal transduction, activating a chain of physiological mechanisms that increase plant tolerance to salt stress [21,22]. Literature shows that rice seedlings boosted salt resistance due to seed priming, and supplemented H_2O_2 application also boosted salinity resistance in rice seedlings [23]. Hemalatha et al. [24] reported that H_2O_2-primed rice seeds had better germination and growth when compared with hydro-primed and unprimed seeds. In addition, hydro-priming has the potential for seed germination and crop growth enhancement. Damalas et al. [25] reported that hydro-priming enhances the germination and field performance of faba bean. Recently, Tania et al. [26] reported that SA, H_2O_2, and hydro-primed wheat seeds enhance germination and seedling growth.

Beans are an important legume crop because of their notable health benefits and soil-friendly properties. The kidney bean is one of the most popular and common varieties of bean and is also called the French bean, snap bean, haricot bean, or navy bean [27]. The seeds of these beans are excellent plant-based sources of protein (23%), flavonoids, and carotenoids [28]. According to the National Nutrient Database of USDA, one standard cup of canned Kidney beans (approximately 150 g) contains 0.55 g of fat, 5.66 g of carbohydrate, 2.6 g of fiber, 1.94 g of sugar, 1.42 g of protein, 17 mg of calcium, 1.2 mg of iron, 18 mg magnesium, 30 mg of phosphorus, and 130 mg potassium [29]. Furthermore, the cultivation of this bean is becoming more popular due to its dual purpose as a pulse while also being consumed as immature tender fruits. A number of studies found that seed priming and the supplementation of various chemicals increased faba bean, broad bean, and common bean performance under stress conditions [17,30,31]. However, research is scant on the germination and seedling growth of kidney beans under stress. To the best of our knowledge, there is no published research available on the role of SA and H_2O_2 priming on the germination and seedling traits of kidney beans under salt stress. Therefore, in this research, we explored the potentiality of SA and H_2O_2 priming on the germination and seedling growth of kidney beans under salt stress.

2. Materials and Methods

2.1. Site of Experiment, Treating Conditions, and Germination Indices Measurement

A Petri-dish and pot experiment was conducted in the laboratory and net house, respectively, at the Seed Science and Technology department, Bangladesh Agricultural University, Mymensingh. The local, high-yielding, popular kidney bean (*Phaseolus vulgaris* L.) variety "Lal Rajma" was collected from the local market (Mymensingh, Bangladesh) and used in the experiment. To prevent the growth of microbial contaminants, present on the seed surface, sodium-hypochlorite (1%) was used for 5 min to sterilize the seeds. The seeds were soaked for 60 min for priming in 1 mM and 2 mM concentrations of SA, and 0.1 mM and 0.15 mM concentrations of H_2O_2, each in individual screw-plugged pots, and in double-distilled water for hydro-priming. Moreover, for control experiments, untreated seeds were used. The concentrations of SA and H_2O_2 were selected based on previous experiments as well as the literature [8,23,26]. Thirty bean seeds were soaked for each concentration of treatment. The primed seeds were positioned on Petri dishes (150 × 25 mm diameter) prepared with 3 layers of Whatman filter papers and kept at room temperature 25 ± 1 °C and relative humidity of 96%. Along with priming, seeds were exposed to salt stress. For the salt treatment, 10 mL of a 150 mM NaCl solution was placed into every Petri dish, while the non-saline condition was controlled by a Petri dish holding 15 mL of pure water. With 3 independent repetitions, the experiment was operated in a completely randomized block design, and the treatment conditions shown in Table 1 were followed. All the chemicals used in this experiment were purchased from Sigma-Aldrich (St. Louis, MO, USA).

Table 1. Treatment conditions and their denotation.

Treatment	Denoted
Control	C
150 mM NaCl	SS
Hydro- primed seed	Hp
Hydro-primed + 150 mM NaCl-treated seed	HpSS
1 mM SA primed seed	SA1
1 mM SA + 150 mM NaCl-treated seed	SA1SS
2 mM SA primed seed	SA2
2 mM SA + 150 mM NaCl-treated seed	SA2SS
0.1 mM H_2O_2 primed seed	HP1
0.1 mM H_2O_2 + 150 mM NaCl-treated seed	HP1SS
0.15 mM H_2O_2 primed seed	HP2
0.15 mM H_2O_2 + 150 mM NaCl-treated seed	HP2SS

Seedling emergence was recorded daily, and radicles 1 mm or more in length were considered germinated. The number of seeds that germinated was documented at 12 h intervals from the first emergence up to the 7th day. After that GP, MGT, and GI were calculated. At 22 days after sowing (DAS), SL (cm) and RL (cm) were measured, and SVI was calculated. According to Rhaman et al. [7], the respective formulas were used to calculate the GP, MGT, GI, and SVI.

$$\text{GP} = \frac{\text{Total number of seeds germinated}}{\text{Total number of seeds placed in germination}} \times 100$$

$$\text{Mean germination time (MGT)} = \sum \frac{Dn}{n}$$

"n" is the seed number on day D and D is the number of days calculated from the beginning of germination.

The GI was computed using the following formula:

$$\text{Germination index (GI)} = \frac{\text{number of seeds germinated}}{\text{day of 1st count}} + \ldots\ldots + \frac{\text{number of seeds germinated}}{\text{day of last count}}$$

$$\text{Seed vigor index (SVI)} = \text{GP} \times \text{seedling length (cm)}$$

where seedling length = shoot length + root length.

2.2. Pot Experiment

Uniform-size seedlings were transplanted from the Petri dish experiments for each treatment and positioned on plastic pots (22.5 × 25.5 cm) packed with soil (6 seedlings/pot; seedling size was approximately 6–7 cm. Though seedlings were collected from Petri dish priming experiments, seedlings (17 days old) were again pretreated with SA and H_2O_2, except hydro-primed and control plants (concentrations are similar to those during the germination period) for 3 days. After that, the seedlings were subjected to 150 mM NaCl stress and grown for 5 days. After 5 days of salt exposure, the plant height, relative water content, and photosynthetic pigments were analyzed. For the fertilizer application, the soil was thoroughly mixed with Hyponex liquid fertilizer (Osaka, Japan), which contains nitrogen (6%), phosphorus (10%), potassium (5%), and other micronutrients. In the pots, 3 mL of the fertilizer solution was applied twice a week in each pot. The treatment conditions were as follows: C, control; SS, 150 mM NaCl; Hp, seedling from hydro-primed seed; HpSS, seedling from hydro-primed seed + 150 mM NaCl; SA1, 1 mM SA; SA1SS, 1 mM SA + 150 mM NaCl; SA2, 2 mM SA; SA2SS, 2 mM SA + 150 mM NaCl; HP1, 0.1 mM H_2O_2; HP1SS, 0.1 mM H_2O_2 + 150 mM NaCl; HP2, 0.15 mM H_2O_2; HP2SS, 0.15 mM H_2O_2 + 150 mM NaCl.

2.3. Relative Water Content Determination

The relative water content (RWC) was finalized using the standard techniques of Mostofa and Fujita [32]. To determine RWC, after 25 days of planting, leaf samples were gathered, and leaves fresh weights (FW) were taken. The leaves were then immersed in dH_2O and remained for 1 h and 2 h, and the turgid weights of leaves were taken. Afterward, surplus water was wiped away from the turgid leaves and the turgid weight (TW) was immediately noted. The leaves were then oven-dried at 70 degrees Celsius for 48 h to determine their dry weight (DW). The RWC was analyzed according to the equation below:

$$\text{RWC (\%)} = (\text{FW} - \text{DW})/(\text{TW} - \text{DW}) \times 100$$

2.4. Estimation of Photosynthetic Pigments

The contents of photosynthetic leaf pigments Chl a and b, carotenoid, lycopene, and beta carotene were quantified using a spectro-photometric approach based on Lichtenthaler's method [33]. Fresh leaves weighing 0.5 g were placed in a tiny vial holding 10 mL of 80% ethanol. For pigment extraction, the bowls were wrapped in aluminum foil and kept in the dark for seven days. Using a spectrophotometer, the absorbance of Chl a and b, lycopene, beta carotene, and carotenoids was determined at 663, 645, 645, 505, and 453 nm wavelengths for chlorophyll, lycopene, beta carotene, and carotenoid concentrations (Shimadzu UV-2550, Kyoto, Japan). The following formulae were used to determine the photosynthetic pigments:

$$\text{Total Chlorophyll} = \text{Chlorophyll a} + \text{Chlorophyll b}$$

$$\text{Chlorophyll a} = (0.999 \times A_{663} - 0.0989 \times A_{645})$$

$$\text{Chlorophyll b} = (-0.328 \times A_{663} + 1.77 \times A_{645})$$

$$\text{Lycopene} = (0.0458 \times A_{663} + 0.204 \times A_{645} + 0.372 \times A_{505} - 0.0806 \times A_{453})$$

$$\text{Beta-carotene} = (0.216 \times A_{663} - 1.22 \times A_{645} - 0.304 \times A_{505} + 0.452 \times A_{453})$$

$$\text{Carotenoids} = (A_{480} + (0.114 \times A_{663} - 0.638 \times A_{645})$$

2.5. Statistical Analysis

Minitab 17.0 statistical software was employed to accomplish a one-way analysis of variance (ANOVA) on the collected data, and Tukey's pairwise comparisons were used to compare statistical variances between the mean values of treatment conditions and salt stress at the 5% significance level. In R 3.6.1, the "pheatmap" was used to create a heatmap and perform hierarchical clustering analysis using Euclidean distances, while the "ggplot2", "factoextra", and FactoMineR" packages were used for principal component analysis (PCA).

3. Results

3.1. Priming Boosts Germination Indices and Traits of Seedlings under Salt Stress

The impacts of SA and H_2O_2 priming on the germination indices of kidney beans under salt stress are displayed in Figure 1. The findings show that salt stress significantly reduced GP by 58.99% compared to the control condition. In both stress and without-stress situations, Hp, SA1, SA2, HP1, and HP2 priming showed a significant effect on GP (Figure 1a). While the highest GP (87.2%) was recorded for SA1, the lowest GP (41.6%) was recorded for HP2SS (Figure 1a). In the case of MGT, significant results were found for different priming treatments compared to salt stress. MGT decreased by 9.6, 29.1, 17.0, 24.8 and 10.5%, respectively, for HpSS, SA1SS, SA2SS, HP1SS, and HP2SS compared to salt stress (Figure 1b). Salt stress significantly reduced GI and priming with SA and H_2O_2 increased GI under salt settings, and the application of priming agents without stress also increased GI (Figure 1c). Similarly, different priming conditions increased SVI but salt stress significantly reduced SVI (66.1%) compared with the control. The results indicated that SVI increased by 119.2, 119.8, 96.9, 81.1 and 62.1%, respectively, for HpSS, SA1SS, SA2SS, HP1SS, and HP2SS priming compared to salt stress (Figure 1d).

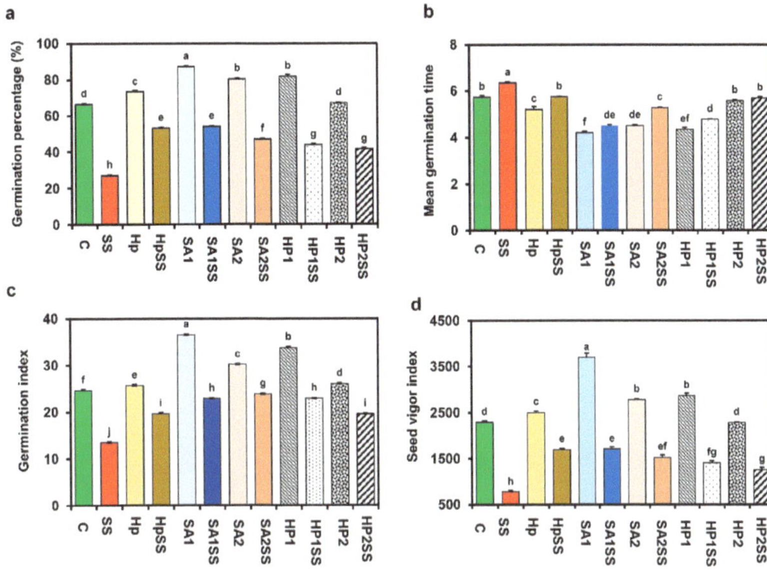

Figure 1. Effects of SA-, H_2O_2-, and hydro-priming on the germination indices of kidney bean under salt stress. (**a**) Germination percentage; (**b**) mean germination time; (**c**) germination index; (**d**) seed vigor index. The error bar represents standard error. Different letters among treatments were analyzed by Tukey's test: $p < 0.05$.

To assess the effects of salt stress and stress-decreasing acts of SA and H_2O_2 on the growth of bean seedlings, we recorded the SL and RL. The results indicated that salt stress significantly diminished SL and RL by 14.8% and 26.6%, respectively, compared with the native plants (Table 2). All the priming conditions significantly increased shoot and root length under salt stress. The highest shoot (30.3 cm) and root (12.0 cm) lengths were observed in the case of SA1 priming. Similarly, shoot fresh weight (SFW), shoot dry weight (SDW), and root fresh and dry weight (RFW, RDW) were decreased by salt stress (Table 2). SA and H_2O_2 priming significantly increased SFW, SDW, RFW, and RDW under salt stress. These results clearly indicate that SA and H_2O_2 priming improved germination indices and seedling traits of kidney beans under salt stress.

Table 2. Effects of SA-, H_2O_2-, and hydro-priming on the seedling traits of kidney bean under salt stress. Shoot length (SL), root length (RL), shoot fresh weight (SFW), shoot dry weight (SDW), root fresh weight (RFW), root dry weight (DRW). Values are means of 3 replicates and means with different letters in a column indicating significant differences ($p \leq 0.05$).

Treatments	SL (cm)	RL (cm)	SFW (gm)	SDW (gm)	RFW (gm)	RDW (gm)
C	27 a	7.5 bcde	4.97 c	0.46 cde	0.577 c	0.07 abcd
SS	23 d	5.5 f	3.27 f	0.303 f	0.323 g	0.03 e
Hp	26.33 bc	7.83 bcd	5.38 b	0.523 bc	0.553 cd	0.06 bcde
HpSS	24.76 bcd	7.23 bcde	4.07 e	0.413 de	0.423 f	0.04 de
SA1	30.33 a	12 a	6.77 a	0.673 a	0.867 a	0.09 a
SA1SS	25.23 bcd	6.26 def	4.30 d	0.426 cde	0.52 cde	0.06 bcde
SA2	26.5 bc	8 bc	5.42 b	0.516 bcd	0.68 b	0.084 ab
SA2SS	25.83 bc	6.5 cdef	4.30 d	0.42 cde	0.48 def	0.05 cde
HP1	26.5 bc	8.5 b	5.5 b	0.61 ab	0.697 b	0.08 abc
HP1SS	25.33 bcd	6.5 cdef	4.2 de	0.37 ef	0.447 ef	0.06 bcde
HP2	26.5 bc	7.5 bcde	5.16 c	0.6 ab	0.58 c	0.072 abcd
HP2SS	24.06 cd	6.1 ef	4.25 de	0.423 cde	0.423 f	0.035 e
SE	0.52	0.48	0.26	0.03	0.04	0.005

3.2. Exogenous SA and H_2O_2 Enhance Growth and RWC of Plants under Salt Stress

The influences of exogenous SA and H_2O_2 on the growth of kidney beans were investigated in this study. When compared to the control, salt stress dramatically reduced plant height (17%) (Figure 2a). Under salt stress, the supplementation of kidney beans with SA (1 mM and 2 mM) and H_2O_2 (0.1 mM and 0.15 mM) improved plant height. Plant height was also increased by a single exposure to SA, H_2O_2, and Hp (Figure 2a).

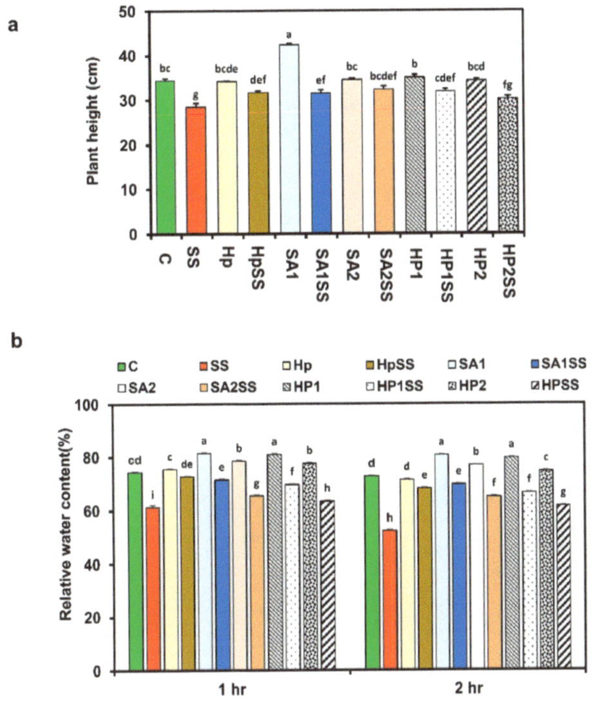

Figure 2. Effects of SA-, H_2O_2-, and hydro-priming on the plant height and relative water content of kidney beans under salt stress. (**a**) Plant height; (**b**) relative water content. The error bar represents the standard error. Different letters among treatments were analyzed by Tukey's test: $p < 0.05$.

The water status of kidney bean plants was studied in this work by measuring RWC with and without salt stress using priming agents. The results showed that salt stress significantly reduced RWC by 17.5% at 1 h and 28.2% at 2 h (Figure 2b). The application of priming agents responded strongly to RWC at both times in comparison to salt conditions. The highest RWC increased by HpSS 15.5% at 1 h, and at 2 h, the maximum RWC displayed by SA1SS was 24.9% higher compared to SS.

3.3. Pretreatment of SA and H_2O_2 Regulate Photosynthetic Pigment of Kidney under Salt Stress

A significant fluctuation of Chl pigment contents was detected due to applied salt stress (Figure 3). A considerable decline in total Chl content (22.3%) including Chl a (23.5%) and Chl b (21%) in bean leaves due to salt stress compared to the control (Figure 3a–c). The supplementation of different concentrations of SA and H_2O_2 remarkably augmented Chl a, Chl b, and total Chl contents. Pigment analysis also revealed that carotenoids (21.8%), lycopene (48.1%), and beta-carotene (53.8%) were also reduced due to salt stress, and the supplementation of SA and H_2O_2 significantly increased the pigments (Figure 3c–f). The supplementation of SA1 under salt stress increased the maximum carotenoids (17.9%), lycopene (52.6%), and beta-carotene (42.6%) contents.

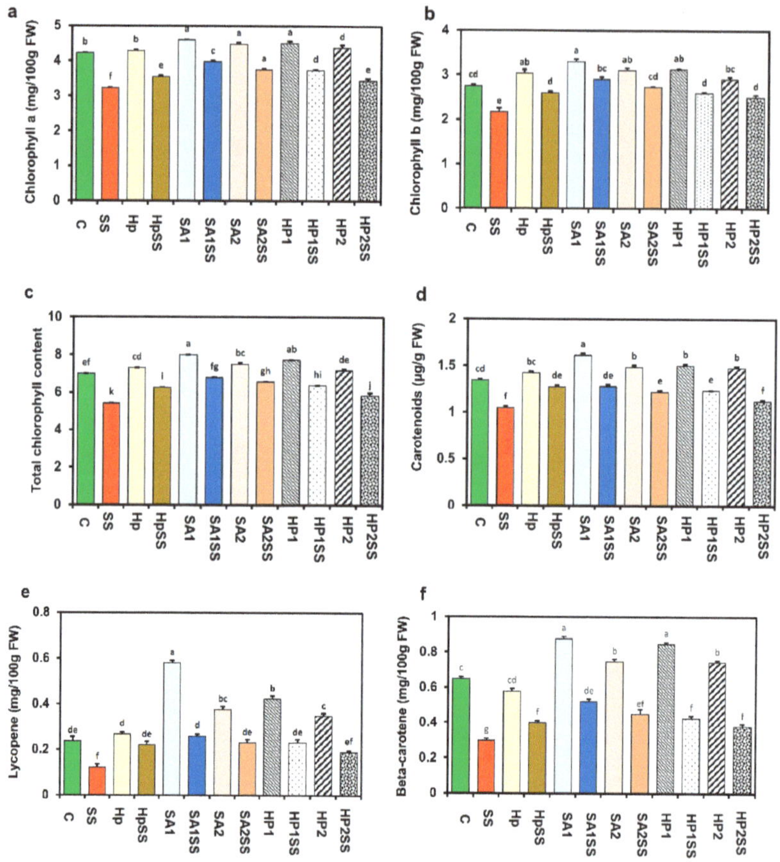

Figure 3. Effects of SA-, H_2O_2-, and hydro-priming on the photosynthetic pigments of kidney bean under salt stress. (**a**) Chlorophyll a; (**b**) chlorophyll a; (**c**) total chlorophyll; (**d**) carotenoids; (**e**) lycopene; (**f**) beta-carotene. The error bar represents the standard error. Different letters among treatments were analyzed by Tukey's test: $p < 0.05$.

3.4. Estimation of Treatment-Variable Interactions through Heatmap and PCA

The mean values of germination indices, seedling traits, and physiological parameters were used to create a heatmap with hierarchical clustering and PCA (Figure 4). Three clusters (Cluster-A, -B, and -C) were found in the variable axis from the hierarchical clustering (Figure 4a). Cluster-A concerns RWC, GI, lycopene (Ly), SFW, SVI, SDW, plant height (PH), SL, and RL. Cluster-A parameters displayed decreasing trends in SS conditions and increasing trends in Hp, HpSS, SA1, SA1SS, SA2, SA2SS, HP1, HP1SS, HP2, and HP2SS conditions. Cluster-B includes the total Chl (TChl), Chl a, Chl b, GP, carotenoid (Car), RWC, beta-carotenoid (Bcar), and RDW. Similar to cluster-A, cluster-B parameters exhibited decreasing trends in SS conditions and increasing trends in other conditions. The variable MGT was clustered in the C group. The PCA components showed PCA 1 (89.8%) and PCA 2 (89.8%) data variability (Figure 4b). The PCA illustrated that cluster-A and B variables are closely associated with different treating conditions, whereas cluster-c was associated with the SS condition.

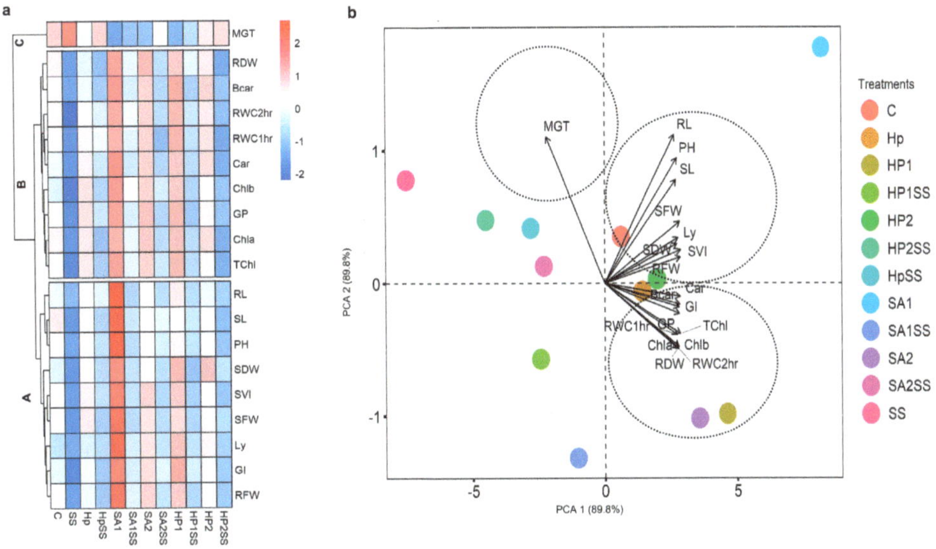

Figure 4. Visualization of the interactions between treatments and all examined factors; a hierarchical clustered heatmap and principal component analysis (PCA) were used. (**a**) Heatmap with clustering method is used to display the scaled average values of all analyzed kidney bean parameters. (**b**) All data were analyzed using PCA. The investigated variables are germination percentage (GP), mean germination time (MGT), germination index (GI), seed vigor index (SVI), shoot length (SL), root length (RL), shoot fresh weight (SFW), shoot dry weight (SDW), root fresh weight (RFW), root dry weight (DRW), lycopene (Ly), plant height (PH), total chlorophyll (TChl), chlorophyll a (Chla), chlorophyll b (Chlb), carotenoid (Car), beta-carotene (Bcar), and relative water content (RWC).

4. Discussion

Successful seed germination is the most significant and fundamental step in the plant growth cycle since it is crucial for seedling development and subsequent productivity [34]. A number of studies described seed priming as a commonly used technique for promoting germination, improving morphological characteristics, and enhancing plant developmental progressions in both non-stress and stress situations [14,35]. Salt stress is a leading form of stress that causes severe reduction in germination and crop establishment. The literature shows that priming and the supplementation of diverse signaling molecules can impart salinity resistance to different crops such as wheat, faba bean, and rice [9,23,26]. Recently, we reported that the priming of wheat seeds with SA and H_2O_2 enhanced germination and seedling parameters under salt stress [26]. The current research was performed to uncover the role of SA and H_2O_2 in the priming of kidney bean seeds under salt stress. The results demonstrated that salt stress considerably decreased GP, GI, and SVI (Figure 1). Salt stress also reduced SL, RL, SFW, SDW, RFW, and RDW (Table 2). The presented results indicated that SA, H_2O_2, and hydro-priming increased GP, GI, SL, RL, and SVI of kidney beans under salt stress. The priming of seeds with these signaling molecules effectively relieved the adverse effects of salt stress on germinating seeds in our current experiment, as evidenced by significantly reduced MGT by SA, H_2O_2, and hydro-priming. These results are consistent with previous research that found that several priming agents significantly reduced the negative effects of salt stress on seed germination-related parameters in wheat [36], maize [37], and rice [38]. Furthermore, heatmap analysis showed that growth traits are increased by priming under salt stress (Figure 4a), and PCA also revealed that seedling traits are closely associated with different priming conditions (Figure 4b). Seed priming with various chemicals may weaken the cell membrane of seeds, resulting in enhanced nutrient utilization capability and faster germination and growth. Additionally, increasing the root length in the present experiment may have allowed for better soil moisture and nutrient uptake, as well as improved overall plant height or growth performance (Figure 2a) [39,40].

Under salt stress, maintaining an adequate water level in plants is a vital physiological step for maintaining normal growth progression [41]. RWC is a well-recognized water status indicator in plants since it is a water-related characteristic [42]. Salinity decreases the water potential in the soil, which has been linked to a decrease in RWC in leaves and a reduction in photosynthesis [43]. Exogenous chemical supplementation has been shown to improve the water status of various plant groups [44–46]. The findings obtained from the study showed RWC declined due to the salt stress (Figure 2b), and this is because the salt-induced damage disrupted the cell wall structure in the leaves, resulting in a decrease in water intake [47]. The results also demonstrated that salt-inhibited RWC is repressed by hydro-, SA-, and H_2O_2-priming and the supplementation of these agents (Figure 2b). This finding indicates that several priming and exogenous substances may be involved in the uptake of additional water from the soil in order to adjust the water level within plant organs.

A common symptom of salt stress is the deterioration of photosynthetic pigments. The present findings showed that salt stress decreased Chl a, Chl b, total Chl, carotenoids, lycopene, and beta carotene, and the priming and exogenous application of different agents improved the photosynthetic pigments under salt stress (Figure 3). This finding was consistent with earlier studies that found that priming rice and wheat seeds with distinct signaling molecules increased photosynthetic pigments [38,48]. The PCA and heatmap also clarified their interaction with treatment agents and stress conditions (Figure 4). These results indicate that different priming agents may aid the photosynthesis system by safeguarding chloroplast pigments from the toxicity of salt, most likely through the oxidative protection of chloroplasts and enhancing the enzyme's activity by regulating chlorophyll biosynthesis [49,50].

Therefore, hydro-, SA-, and H_2O_2-priming enhance the germination and seedling traits of kidney beans under salt stress. Though all agents were determined to be viable options for improving kidney bean salinity tolerance, the lowest concentration of SA was found to be a more promising candidate for the improvement of salt-inhibited germination and seedling traits of kidney beans under salt stress.

5. Conclusions

The findings concluded that salt stress reduces germination indices, growth traits, leaf water status, and photosynthetic pigments of kidney beans. The priming and exogenous application of SA and H_2O_2 enhance the germination percentage, germination index, seed vigor index, shoot and root length, leaf water status, and photosynthetic pigments of kidney beans under salt stress. Our results also suggest that lower concentrations of SA and H_2O_2 can be used to achieve success in kidney bean cultivation. In spite of this, it is recommended that future research conducts a trial at the field level to validate our findings.

Author Contributions: Conceptualization, M.S.R.; methodology, M.S.R.; software, S.S.T.; validation, M.S.R., M.A.H., and Y.M.; formal analysis, S.S.T.; investigation, S.S.T., F.R., and M.M.R., M.H.K.; writing—original draft preparation, S.S.T. and F.R.; writing—review and editing, M.S.R. and Y.M.; visualization, M.S.R., M.A.H., and Y.M.; supervision, M.S.R. All authors have read and agreed to the published version of the manuscript.

Funding: The present investigation did not receive any external funds.

Institutional Review Board Statement: Not applicable.

Informed Consent Statement: Not applicable.

Data Availability Statement: Not applicable.

Conflicts of Interest: The authors declare no conflict of interest.

References

1. Finch-Savage, W.E.; Bassel, G.W. Seed vigour and crop establishment: Extending performance beyond adaptation. *J. Exp. Bot.* **2016**, *67*, 567–591. [CrossRef] [PubMed]
2. Rhaman, M.S.; Kibria, M.G.; Hoque, A. Climate Change and Its Adverse Impacts on Plant Growth in South Asia: Current Status and Upcoming Challenges. *Phyton* **2022**, *91*, 695.
3. Arun, M.N.; Hebbar, S.S.; Senthivel, T.; Nair, A.K.; Padmavathi, G.; Pandey, P.; Singh, A. Seed Priming: The Way Forward to Mitigate Abiotic Stress in Crops. In Plant Stress Physiology-Perspectives in Agriculture. In *Plant Stress Physiology-Perspectives in Agriculture*; IntechOpen: London, UK, 2022. [CrossRef]
4. Isayenkov, S.V.; Maathuis, F.J. Plant salinity stress: Many unanswered questions remain. *Front. Plant Sci.* **2019**, *10*, 80. [CrossRef] [PubMed]
5. Gopalakrishnan, T.; Kumar, L. Modeling and mapping of soil salinity and its impact on Paddy Lands in Jaffna Peninsula, Sri Lanka. *Sustainability* **2020**, *12*, 8317. [CrossRef]
6. Machado, R.M.A.; Serralheiro, R.P. Soil Salinity: Effect on Vegetable Crop Growth. Management Practices to Prevent and Mitigate Soil Salinization. *Horticulturae* **2017**, *3*, 30. [CrossRef]
7. Rhaman, M.S.; Kibria, M.G.; Hossain, M.; Hoque, M.A. Effects of organic manure and bio-slurries with chemical fertilizers on growth and yield of rice (cv. BRRI dhan28). *Int. J. Expt. Agric.* **2016**, *6*, 36–42.
8. Rhaman, M.S.; Rauf, F.; Tania, S.S.; Karim, M.M.; Sagar, A.; Robin, A.H.K.; Murata, Y. Seed priming and exogenous application of salicylic acid enhance growth and productivity of okra (*Abelmoschus esculentus* L.) by regulating photosynthetic attributes. *J. Exp. Biol. Agric. Sci.* **2021**, *9*, 759–769. [CrossRef]
9. Abdel Latef, A.A.H.; Tahjib-Ul-Arif, M.; Rhaman, M.S. Exogenous auxin-mediated salt stress alleviation in faba bean (*Vicia faba* L.). *Agronomy* **2021**, *11*, 547. [CrossRef]
10. Rasel, M.; Tahjib-Ul-Arif, M.; Hossain, M.A.; Hassan, L.; Farzana, S.; Brestic, M. Screening of salt-tolerant rice landraces by seedling stage phenotyping and dissecting biochemical determinants of tolerance mechanism. *J. Plant Growth Regul.* **2021**, *40*, 1853–1868. [CrossRef]
11. Tania, S.S.; Rhaman, M.S.; Hossain, M.M. Hydro-priming and halo-priming improve seed germination, yield and yield contributing characters of okra (*Abelmoschus esculentus* L.). *Trop. Plant Res.* **2020**, *7*, 86–93. [CrossRef]
12. Rhaman, M.S.; Rauf, F.; Tania, S.S.; Khatun, M. Seed priming methods: Application in field crops and future perspectives. *Asian J. Res. Crop Sci.* **2020**, *5*, 8–19. [CrossRef]
13. Zhu, Z.H.; Sami, A.; Xu, Q.Q.; Wu, L.L.; Zheng, W.Y.; Chen, Z.P.; Jin, X.Z.; Zhang, H.; Li, Y.; Yu, Y.; et al. Effects of seed priming treatments on the germination and development of two rapeseed (*Brassica napus* L.) varieties under the co-influence of low temperature and drought. *PLoS ONE* **2021**, *16*, e0257236. [CrossRef] [PubMed]
14. Rhaman, M.S.; Imran, S.; Rauf, F.; Khatun, M.; Baskin, C.C.; Murata, Y.; Hasanuzzaman, M. Seed priming with phytohormones: An effective approach for the mitigation of abiotic stress. *Plants* **2021**, *10*, 37. [CrossRef] [PubMed]
15. Pastor, V.; Luna, E.; Mauch-Mani, B.; Ton, J.; Flors, V. Primed plants do not forget. *Environ. Exp. Bot.* **2013**, *94*, 46–56. [CrossRef]
16. Borges, A.A.; Jiménez-Arias, D.; Expósito-Rodríguez, M.; Sandalio, L.M.; Pérez, J.A. Priming crops against biotic and abiotic stresses: MSB as a tool for studying mechanisms. *Front. Plant. Sci.* **2014**, *5*, 642. [CrossRef] [PubMed]
17. Azooz, M.M. Salt stress mitigation by seed priming with salicylic acid in two faba bean genotypes differing in salt tolerance. *Int. J. Agric. Biol.* **2009**, *11*, 343–350.
18. Roy, P.R.; Tahjib-Ul-Arif, M.; Polash, M.A.; Hossen, M.; Hossain, M.A. Physiological mechanisms of exogenous calcium on alleviating salinity-induced stress in rice (*Oryza sativa* L.). *Physiol. Mol. Biol. Plants* **2019**, *25*, 611–624. [CrossRef]
19. Khan, W.; Prithiviraj, B.; Smith, D.L. Photosynthetic responses of corn and soybean to foliar application of salicylates. *J. Plant Physiol.* **2003**, *160*, 485–492. [CrossRef]
20. Koo, Y.M.; Heo, A.Y.; Choi, H.W. Salicylic acid as a safe plant protector and growth regulator. *Plant Pathol. J.* **2020**, *36*, 1. [CrossRef]
21. Uchida, A.; Jagendorf, A.T.; Hibino, T.; Takabe, T. Effects of hydrogen peroxide and nitric oxide on both salt and heat stress tolerance in rice. *Plant Sci.* **2002**, *163*, 515–523. [CrossRef]
22. Silva, P.C.; AzevedoNeto, A.D.; Gheyi, H.R.; Ribas, R.F.; Silva, C.R.; Cova, A.M. Seed priming with H_2O_2 improves photosynthetic efficiency and biomass production in sunflower plants under salt stress. *Arid. Land Res. Manag.* **2021**, *5*, 1–5. [CrossRef]
23. Roy, P.R.; Tahjib-Ul-Arif, M.; Akter, T.; Ray, S.R.; Sayed, M.A. Exogenous ascorbic acid and hydrogen peroxide alleviates salt-induced oxidative stress in rice (*Oryza sativa* L.) by enhancing antioxidant enzyme activities and proline content. *Adv. Environ. Biol.* **2016**, *10*, 148–155.
24. Hemalatha, G.; Renugadevi, J.; Eevera, T. Studies on seed priming with hydrogen peroxide for mitigating salt stress in rice. *Int. J. Curr. Microbiol. App. Sci.* **2017**, *6*, 691–695. [CrossRef]
25. Damalas, C.A.; Koutroubas, S.D.; Fotiadis, S. Hydro-priming effects on seed germination and field performance of faba bean in spring sowing. *Agriculture* **2019**, *9*, 201. [CrossRef]
26. Tania, S.S.; Rahaman, M.; Rauf, F.; Afroj Suborna, M.; Humayun Kabir, M.; Hoque, A.; Rhaman, M.S. Seed priming with Salicylic Acid (SA) and Hydrogen Peroxide (H_2O_2) Improve Germination and Seedling Growth of Wheat (*Triticum aestivum* L.) under Salt Stress. *Asian J. Res. Crop Sci.* **2021**, *6*, 60–69. [CrossRef]
27. Fatema, R.; Rahman, J.; Shozib, H.B.; Nazrul, M.I.; Fatima, K. Genetic diversity and nutritional components evaluation of Bangladeshi germplasms of kidney bean (*Phaseolus vulgaris* L.). *J. Genet. Res.* **2019**, *5*, 83–96.

28. Mena, E.; Leiva-Mora, M.; Jayawardana, E.K.; García, L.; Veitía, N.; Bermúdez-Caraballoso, I.; Collado, R.; Ortíz, R.C. Effect of salt stress on seed germination and seedlings growth of *Phaseolus vulgaris* L. *Cult. Trop.* **2015**, *36*, 71–74.
29. USDA (National Nutrient Database). Cut Green Beans. 29 August 2017. Available online: https://ndb.nal.usda.gov/ (accessed on 1 April 2022).
30. Azooz, M.M.; Alzahrani, A.M.; Youssef, M.M. The potential role of seed priming with ascorbic acid and nicotinamide and their interactions to enhance salt tolerance in broad bean (*'Vicia faba'L.*). *Aust. J. Crop Sci.* **2013**, *7*, 2091–2100.
31. Nouairi, I.; Jalali, K.; Zribi, F.; Barhoumi, F.; Zribi, K.; Mhadhbi, H. Seed priming with calcium chloride improves the photosynthesis performance of faba bean plants subjected to cadmium stress. *Photosynthetica* **2019**, *57*, 438–445. [CrossRef]
32. Mostofa, M.G.; Fujita, M. Salicylic acid alleviates copper toxicity in rice (*Oryza sativa* L.) seedlings by up-regulating antioxidative and glyoxalase systems. *Ecotoxicology* **2013**, *22*, 959–973. [CrossRef]
33. Lichtenthaler, H.K. Chlorophylls and carotenoids: Pigments of photosynthetic bio-membranes. *Meth. Enzymol.* **1987**, *148*, 350–382.
34. Shiade, S.R.; Boelt, B. Seed germination and seedling growth parameters in nine tall fescue varieties under salinity stress. *Acta Agric. Scand.-B Soil Plant Sci.* **2020**, *17*, 485–494. [CrossRef]
35. Muhei, S.H. Seed priming with phytohormones to improve germination under dormant and abiotic stress conditions. *Adv. Crop Sci. Technol.* **2018**, *6*, 403–409. [CrossRef]
36. Farooq, M.; Basra, S.M.A.; Rehman, H.; Saleem, B.A. Seed priming enhances the performance of late sown wheat (*Triticum aestivum* L.) by improving chilling tolerance. *J. Agron. Crop. Sci.* **2008**, *194*, 55–60. [CrossRef]
37. Ashraf, M.; Rauf, H. Inducing salt tolerance in maize (*Zea mays* L.) through seed priming with chloride salts: Growth and ion transport at early growth stages. *Acta Physiol. Plant.* **2001**, *23*, 407–414. [CrossRef]
38. Tahjib-Ul-Arif, M.; Afrin, S.; Polash, M.A.S.; Akter, T.; Ray, S.R.; Hossain, M.; Hossain, M.A. Role of exogenous signaling molecules in alleviating salt-induced oxidative stress in rice (*Oryza sativa* L.): A comparative study. *Acta Physiol. Plant.* **2019**, *41*, 1–14. [CrossRef]
39. Muhammad, I.; Kolla, M.; Volker, R.; Günter, N. Impact of nutrient seed priming on germination, seedling development, nutritional status and grain yield of maize. *J. Plant Nutr.* **2015**, *38*, 1803–1821. [CrossRef]
40. Adhikari, B.; Dhital, P.R.; Ranabhat, S.; Poudel, H. Effect of seed hydro-priming durations on germination and seedling growth of bitter gourd (*Momordica charantia*). *PLoS ONE.* **2021**, *16*, e0255258. [CrossRef]
41. Ahanger, M.A.; Agarwal, R.M. Salinity stress induced alterations in antioxidant metabolism and nitrogen assimilation in wheat (*Triticum aestivum* L.) as influenced by potassium supplementation. *Plant Physiol. Biochem.* **2017**, *115*, 449–460. [CrossRef]
42. Li, Y.; Li, H.; Li, Y.; Zhang, S. Improving water-use efficiency by decreasing stomatal conductance and transpiration rate to maintain higher ear photosynthetic rate in drought-resistant wheat. *Crop J.* **2017**, *5*, 231–239. [CrossRef]
43. Pattanagul, W.; Thitisaksakul, M. Effect of salinity stress on growth and carbohydrate metabolism in three rice (*Oryza sativa* L.) cultivars differing in salinity tolerance. *Indian J. Exp. Biol.* **2008**, *46*, 736–742. [PubMed]
44. Saboon, R.I.; Ahmad, N.; Ilyas, N.; Batool, N.; Gul, S. Salicylic acid enhances wheat plant growth under water stress conditions. *Int. J. Biol. Biotech.* **2015**, *12*, 329–336.
45. Tahjib-Ul-Arif, M.; Siddiqui, M.N.; Sohag, A.A.; Sakil, M.A.; Rahman, M.M.; Polash, M.A.; Mostofa, M.G.; Tran, L.S. Salicylic acid-mediated enhancement of photosynthesis attributes and antioxidant capacity contributes to yield improvement of maize plants under salt stress. *J. Plant Growth Regul.* **2018**, *37*, 1318–1330. [CrossRef]
46. Hernández, J.A.; Díaz-Vivancos, P.; Acosta-Motos, J.R.; Barba-Espín, G. Potassium nitrate treatment is associated with modulation of seed water uptake, Antioxidative Metabolism and Phytohormone Levels of Pea Seedlings. *Seeds* **2022**, *1*, 5–15. [CrossRef]
47. Parvin, K.; Hasanuzzaman, M.; Bhuyan, M.H.; Nahar, K.; Mohsin, S.M.; Fujita, M. Comparative physiological and biochemical changes in tomato (*Solanum lycopersicum* L.) under salt stress and recovery: Role of antioxidant defense and glyoxalase systems. *Antioxidants* **2019**, *8*, 350. [CrossRef]
48. Farzana, S.; Rasel, M.; Tahjib Ul Arif, M.; Hossain, M.A.; Azam, M.G.; Al Galib, M.A.; Mahamud, A.S.U.; Hossain, M.A. Salicylic acid and thiourea ameliorate the negative impact of salt stress in wheat (*Triticum aestivum* L.) seedlings by up-regulating photosynthetic pigments, leaf water status, and antioxidant defense system. *J. Phytol.* **2021**, *13*, 130–145. [CrossRef]
49. Amin, A.A.; AEK, A.A.; Abouziena, H.F.; El-Awadi, M.; Gharib, F.A. Effects of benzoic acid and thiourea on growth and productivity of wheat (*Triticum aestivum* L.) plants. *Int. Sci. Res. J.* **2016**, *72*, 132–149.
50. Ma, X.; Zheng, J.; Zhang, X.; Hu, Q.; Qian, R. Salicylic acid alleviates the adverse effects of salt stress on *Dianthus superbus* (Caryophyllaceae) by activating photosynthesis, protecting morphological structure, and enhancing the antioxidant system. *Front. Plant Sci.* **2017**, *8*, 600. [CrossRef]

seeds

MDPI

Article

Effects of Chemical Priming on the Germination of the Ornamental Halophyte *Lobularia maritima* under NaCl Salinity

Iman Zammali, Amira Dabbous, Seria Youssef and Karim Ben Hamed *

Laboratory of Extremophile Plants, Centre of Biotechnology of Borj Cedria, P.O. Box 901, Hammam Lif 2050, Tunisia; imanzammeli9@gmail.com (I.Z.); amiradabousbenali@gmail.com (A.D.); seriayoussef93@gmaill.com (S.Y.)
* Correspondence: karim.benhamed@cbbc.rnrt.tn

Abstract: *Lobularia maritima* or sweet alyssum (Brassicaceae) is an annual littoral halophyte, naturally thriving on sandy beaches. In addition to its obvious interest as a naturally salt-tolerant plant, this species is mainly cultivated as an ornamental plant in many countries. Laboratory experiments were carried out to assess the effects of salinity on seed germination and on germination recovery from the effects of saline conditions after transfer to distilled water. Seed germination responses were determined at 0, 50, 100, 200 and 300 mM NaCl. Salt (NaCl) does not affect the germination of *L. maritima* if applied at a moderate dose of 50 mM. For higher concentrations of NaCl, there is a decrease in the germination rate at 100 and 200 mM NaCl or even a total inhibition of germination at 300 mM. Salt lowers or inhibits germination only through osmotic effects. To improve the germination of *L. maritima* under high salinity, seeds were pretreated with KNO_3, thiourea, proline and salicylic acid. The germination of seeds is improved by KNO_3 in the presence or absence of salt, while thiourea increases the final germination without affecting the germination rate. Salicylic acid amplifies the effect of salt, while proline delays germination without stopping it completely. These findings indicate that the application of KNO_3 and thiourea may be used to improve seed germination of *L. maritima*, which is of great interest for cultivating this plant for landscaping purposes in saline soils.

Keywords: germination; sweet alyssum; landscaping; ornamental; seed priming; salt stress

Citation: Zammali, I.; Dabbous, A.; Youssef, S.; Ben Hamed, K. Effects of Chemical Priming on the Germination of the Ornamental Halophyte *Lobularia maritima* under NaCl Salinity. *Seeds* **2022**, *1*, 99–109. https://doi.org/10.3390/seeds1020009

Academic Editors: José Antonio Hernández Cortés, Gregorio Barba-Espín and Pedro Diaz-Vivancos

Received: 7 March 2022
Accepted: 18 April 2022
Published: 25 April 2022

Publisher's Note: MDPI stays neutral with regard to jurisdictional claims in published maps and institutional affiliations.

1. Introduction

In recent years, rapid urbanization has increased the demand for landscaping in many countries. At the same time, the availability of good-quality water for gardens and landscapes is becoming increasingly restricted due to the rising demand for domestic use of the scarce fresh water resources. Consequently, landscape architects are in search of ornamental plants that perform well with saline groundwater and treated wastewater. To date, only limited systematic work has been carried out to study the tolerance of landscaping plants to higher levels of salinity. Ornamental flowering plants, such as carnations, roses, chrysanthemums, and gerberas, found around the world, albeit known for their high economic, ornamental, ecological, and edible value, are sensitive to salt stress. Recent studies showed that the salt tolerance of ornamental crops such as chrysanthemum can be improved through genetic manipulation [1,2].

Alternatively, some naturally salt-tolerant plants (halophytes), such as *Aster tripolium*, *Sesuvium portulacastrum*, *Limonium* sp., *Crithmum maritimum* and *Lobularia maritima*, can be used for landscaping and ornamental purposes, and cultivated in salt-affected areas under salinity irrigation [3]. According to the e-HALOPH database [4], many halophyte species produce attractive flowers. Cassaniti and Romano [5] investigated halophytes native to the Mediterranean region and listed 13 families with about 42 species of ornamental potential. Although many halophytes can be considered as promising ornamental crops, only little information is available in the literature regarding their cultivation techniques.

Salinity is known to impair seed germination of halophyte species and they may be as salt-sensitive as glycophytes in the early stages of their life cycle [6] Under natural conditions, seeds of halophytes are subjected to salt stress dominated by NaCl, yet, other salts can also affect seeds germination significantly due to the low osmotic potentials occurring under saline conditions or to the toxic effects of the ions [6]. Salinity induces both a reduction in the percentage of seeds germinating and a delay in the initiation of the germination process. At salinities beyond the tolerance limits of the species, complete inhibition of the germination process can occur because of: (i) inadequate imbibition; (ii) ionic toxicity; (iii) interference with metabolism; (iv) destruction of enzymes; and (v) imbalance of growth regulators [7]. Once salinity is reduced following rain, seeds of halophytes have the capacity to recover from the salinity shock and start germination for non-dormant seeds [8]. Impaired germination is partly resulting from the salt-induced reduction of the growth regulator contents in seeds, which control germination activity in most halophyte seeds [8,9]. Interestingly, seed dormancy can be broken either by physical treatments, including scarification, or by chemical treatments by the exogenous application of gibberellins, nitrogen or ethanol, which can efficiently alleviate the harmful effect of salinity on halophyte germination [10–13].

Lobularia maritima L. Desv., commonly known as sweet alyssum—is a species of the Brassicaceae that is distributed throughout the Mediterranean basin, where it grows in coastal zones, dunes and scrublands. This plant forms a basal rosette with prostrated-ascending flowering stems. The plant is a short-lived perennial with a lifespan of approximately three years but with a long flowering period because *L. maritima* individuals produce flowering stems more or less regularly over the completely flowering season [14]. It has been cultivated both as an ornamental plant in many countries in the world and as an insectary plant to intercrop with lettuce because alyssum flowered quickly after planting was not overly aggressive or likely to become a weed, and attracted several beneficial species, but few pest species [15,16].

In the current study, our aim was to address the following three questions: (1) is germination of *L. maritima* inhibited by an osmotic effect and/or a specific ion effect, (2) to what extent can seeds recover from exposure to high concentrations of sodium chloride, and (3) which suitable (pre)treatment that improves germination and overcomes the effects of NaCl by investigating the role of KNO_3, thiourea, proline and salicylic acid during the germination of *L. maritima* under increasing NaCl salinity in order to get more information on the mechanisms by which salinity may inhibit germination of this species in the natural conditions.

2. Materials and Methods

2.1. Plant Materials

Seeds of *L. maritima* were harvested on May 2019 in Tabarka, a locality close to the Mediterranean seashore, NE Tunisia (36°57′23″ N 8°45′28.5″ E). They were separated from the inflorescences and stored dry at 4 °C. They were sterilized using 70% ethanol for 5 min, 40% bleach for 10 min and washed five times with sterile water.

2.2. Germination Tests

Seeds were placed in tightly sealed 9 cm Petri dishes with Whatman paper # 1 moistened in 2 mL of distilled water or a test saline solution. Three replicates of 20 seeds were used for each treatment. Germination was carried out in the dark at 20 °C (incubator KRG-250 model). The dishes were checked daily (for 45 days) to count the germinated seeds. Seeds developing radicals of >1 mm length were considered germinated.

For evaluation of salinity tolerance at seedling stage, seeds were sown in Petri dishes on filter paper soaked with NaCl solution (50, 100, 200 and 300 mM). Control seeds were treated continuously with distilled water. Ungerminated salt-treated seeds were transferred to filter paper soaked with distilled water for another 7 days to determine their ability to recover from salt pretreatments. To improve *L. maritima* seed germination under increasing NaCl concentration, seeds were soaked for 24 h in four chemical treatments, afterwards, they were washed thoroughly with distilled water, dried (to obtain the initial seed weight), and kept for germination at 25 °C in growth chamber. The impact of the external application of KNO_3 (10 mM), thiourea (50 mM), proline (1 mM) or salicylic acid (10 µM) on germination was tested over a range of NaCl concentrations (0, 50, 100, 200 and 300 mM). These concentrations of priming substances were selected on the basis of a preliminary trial (data not given).

2.3. Data Analyses

Three parameters of germination were determined: final germination percentage, germination rate and germination recovery percentage. For each Petri dish, final germination percentage was calculated as:

$$\text{Final germination} = (\text{number of germinated seeds}/\text{number of sampled seeds}) \times 100.$$

The rate of germination was estimated using a modified Timson's index of germination velocity:

$$\text{Rate of germination} = \Sigma\, G/t$$

where G is the percentage of seed germination at each measurement and t is the number of measurements.

The germination recovery percentage was determined by the following formula:

$$\text{Germination recovery} = [(a - b)/(c - b)] \times 100$$

where a is the number of seeds germinated after being transferred to distilled water, b is the number of seeds germinated in saline solution, and c is the total number of seeds.

The data were subjected to one-way analysis of variance (ANOVA) to evaluate the effect of the salinity and their interactions on the final germination percentage, germination rate and recovery percentage. Tukey's test (honestly significant differences, HSD) was carried out to perform all-pairwise comparisons between individual treatments ($p < 0.05$). All statistical analyses were performed using SPSS Statistics 20 (IBM Corp., Armonk, NY, USA) software.

3. Results

3.1. Effects of Salt Treatment on Final Germination

The kinetics of germination at 50 mM NaCl are generally similar to those observed in the control medium (distilled water), except that the germination rate is higher in the presence than in the absence of salt (Figure 1). Germination does not exceed 20% at 100 mM and 200 mM NaCl. It is completely inhibited at 300 mM NaCl (Figure 1).

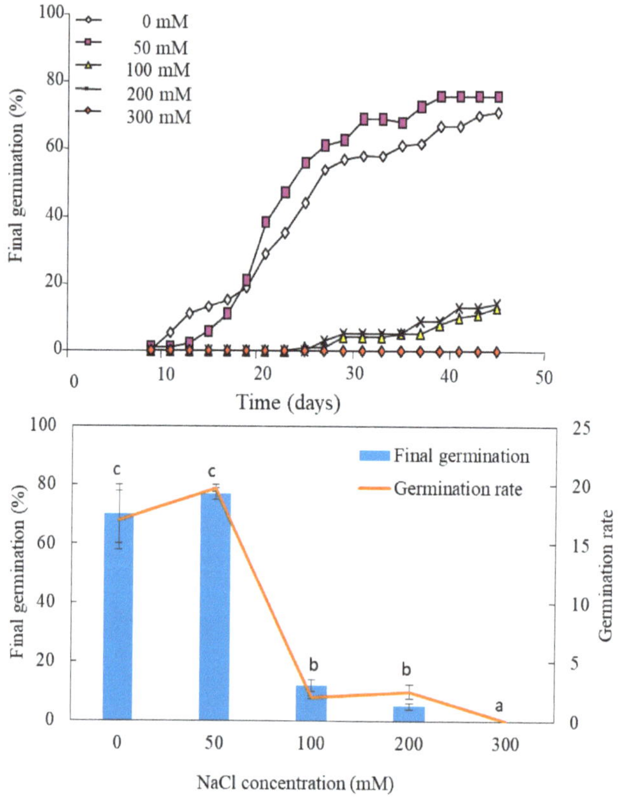

Figure 1. Effect of salinity on the germination of *Lobularia maritima*. Values at each level of salinity with the same letter are not significantly different ($p < 0.05$); Tukey test. Data are presented as means and standard errors of three repetitions.

3.2. Effects of Salt Treatment on Germination Recovery

The test of the reversibility of germination consists in transferring the seeds that have not germinated on salt medium to a control medium devoid of NaCl in order to determine the nature of the effect of salt: toxic or osmotic. Seeds that have not germinated at 100, 200 or 300 mM, once transferred to distilled water, regain their ability to germinate at rates of 80%, 65% and 50%, respectively (Figure 2). As the germination capacity of the seeds that have not undergone any treatment is of the order of 80%, it is deduced that the stay of the seeds for 45 days in the presence of salt did not affect their viability. The inhibition of germination previously observed in the presence of salt is therefore mainly of an osmotic nature. However, a toxic effect is also likely at 100, 200 and 300 mM, since the seeds do not regain their maximum germination capacity. The body of data shows that salt inhibits the germination of *L. maritima* seeds mainly by an osmotic effect.

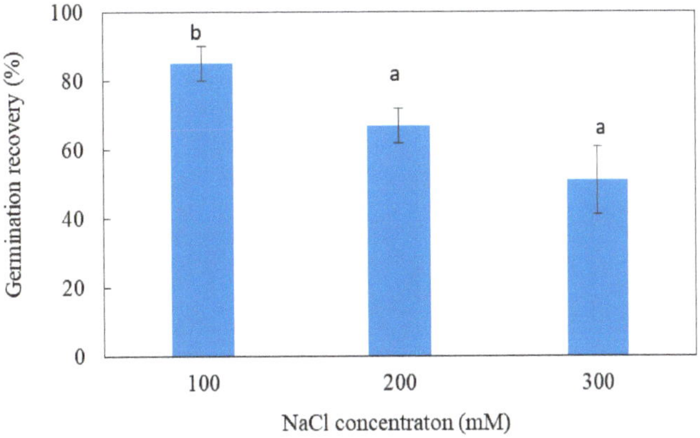

Figure 2. Germination recovery of ungerminated salt-treated seeds of *L. maritima*. Values at each level of salinity with the same letter are not significantly different ($p < 0.05$); Tukey test. Data are presented as means and standard errors of three repetitions.

3.3. *Effects of Chemical Priming on Germination under NaCl Salinity*

In the absence of salt, nitrate treatment accelerates germination (Figure 3A). Thus, after 20 days, the germination rate reaches 60% in the presence of nitrate against only 30% in its absence (Figure 4). Nitrogen treatment also increases the final germination: this parameter increases from 70% on control medium to 82% when nitrate is provided in the imbibition solution. The beneficial effect of nitrate on both components of germination (rate and final germination capacity) also appears at 50 mM NaCl, but this effect is much less pronounced than before. The beneficial effect of nitrate appears more significantly at 100 and 200 mM NaCl. This effect is exerted in a more pronounced way on the percentage than on the rate of germination. Thus, at these doses of salt, the first germinations are observed between 20 and 25 days in the presence of NO_3^- against 30 days in the absence of NO_3^-. The final percentage of germination reaches 78% and 38%, respectively, at 100 and 200 mM for the first treatment. The beneficial action of NO_3^- also appears at 300 mM NaCl since 10% of the seeds germinate against a zero germination rate in the absence of this nitrogenous source. In conclusion, NO_3^- improves the germination capacity of *L. maritima* seeds, especially in the presence of salt.

Monitoring of germination kinetics in the presence of thiourea shows that the effect of this substance is dependent on the salinity of the medium (Figure 3B). The application of thiourea in the absence of salt does not significantly change the kinetics of germination. At 50 mM NaCl, the intake of thiourea slows down germination: the maximum germination rate is observed on the 35th day in the absence of this substance and on the 45th day in its presence. The final percentage of germination under salinity is also affected (Figure 4). At 100, 200 and 300 mM, thiourea improves germination capacity, which is manifested mainly by a shortening of the latency phase and particularly by an increase in germination capacity. This effect appears particularly at 200 mM NaCl, the final germination increases from about 5% in the absence of thiourea to 17% in its presence. There is also a partial lifting of dormancy induced by 300 mM NaCl. Overall, the effect of thiourea is reminiscent of that of NO_3^- with an acceleration of germination and an increase in the percentage of sprouted seeds, particularly at 100 and to a lesser degree at 200 mM. Nevertheless, this beneficial action remains less pronounced than that of KNO_3.

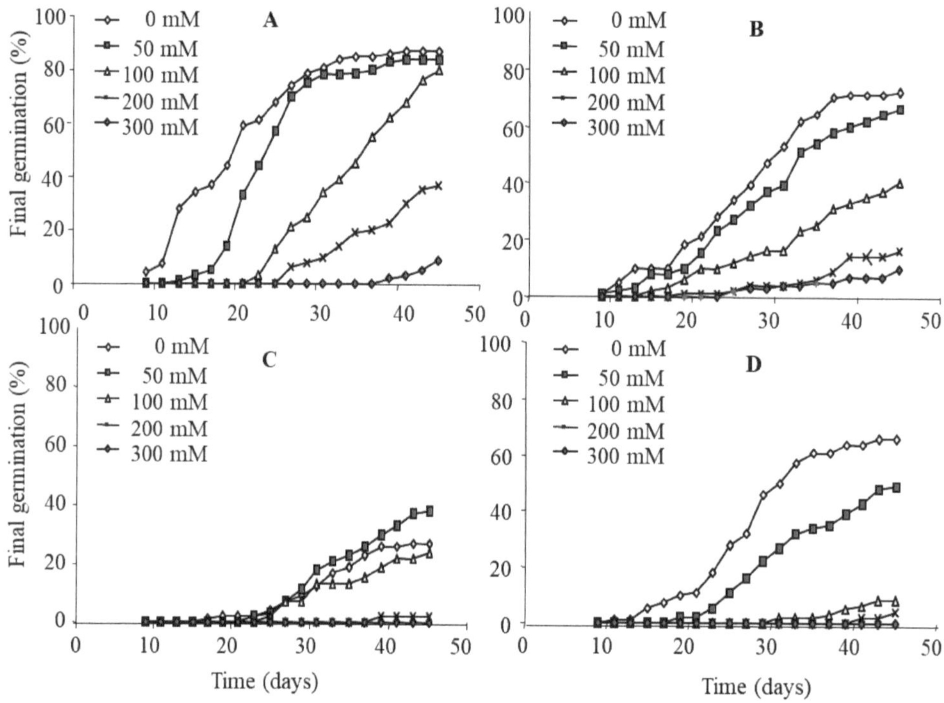

Figure 3. Effect of KNO_3 (**A**), thiourea (**B**), proline (**C**) and salicylic acid (**D**) pretreatment on the germination of *L. maritima*.

The application of proline at a dose of 1 mM lowers the germination rate and dramatically delays the germination process (Figure 3C). This action is noted for all doses of salt applied. Thus, after 45 days, the germination capacity reaches 27% in the absence of salt. Seeds subjected to 50 and 100 mM NaCl express a behavior close to that observed in the absence of salt. Since the germination process is still underway even after 45 days of germination, it is difficult to diagnose the nature of the depressive action of proline: is it a toxic or osmotic effect? The second type of effect would be unlikely since proline is given at a dose of 1 mM which will not significantly change the concentration of the imbibition solution which contains up to 200 mM NaCl. Moreover, we did not record an accentuation of the osmotic effect of salt by the intake of 20 mM of KNO_3 or 10 mM of thiourea. Given the unexpected effect of proline, several trials have been conducted and have confirmed the depressive action of this amino acid on germination.

In the absence of salt, salicylic acid does not appear to alter the kinetics of germination (Figure 3D). However, it accentuates the depressive action of salt. Indeed, we note a cancellation of the stimulatory effect of 50 mM NaCl with an increase in the duration of the latency phase and a decrease in the final germination from 77% in the absence of salicylic acid to 42% in its presence. At 100 and 200 mM NaCl, germination rates are also lower in the presence than in the absence of salicylic acid (Figure 4).

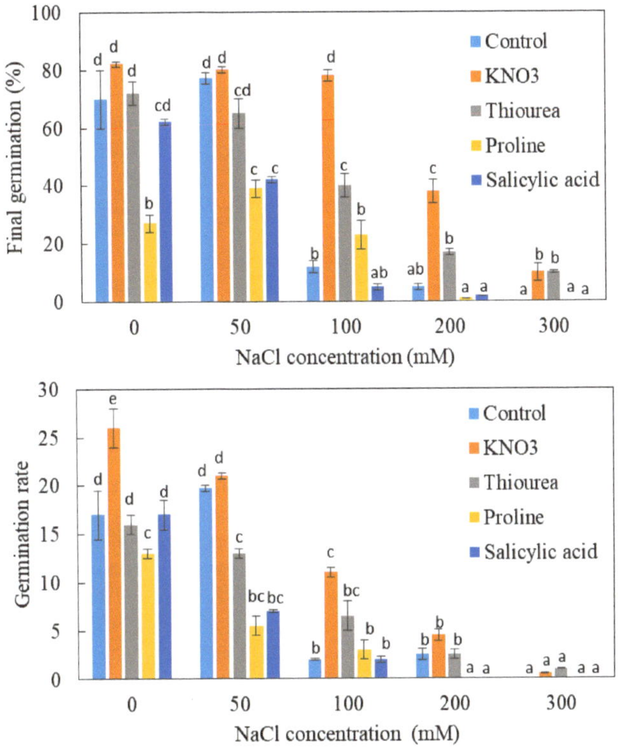

Figure 4. Comparative analysis of the final germination and germination rate of the unprimed (control) and primed seeds of *L. maritima*. Values at each level of salinity with the same letter are not significantly different ($p < 0.05$); Tukey test. Data are presented as means and standard errors of three repetitions.

4. Discussion

Salt (NaCl) does not affect the germination of *L. maritima* if applied at a moderate dose of 50 mM. Nevertheless, for higher concentrations of NaCl, there is a decrease in the germination rate or even a total inhibition of germination (200 and 300 mM). The depressive effect of NaCl on final germination and germination rate has been demonstrated in several halophytes. However, these plants maintain an ability to germinate (even at reduced rates) in a wide range of saline concentrations. Thus, the tolerance limits for salinity at the germination stage vary from one species to another. For example, the germination of *Urochondra setulosa* shows a considerable decrease of 80% at 300 mM NaCl [17]. The seeds of *Suaeda moquinii*, *Salicornia rubra*, *Halogeton glomeratus* and *Kochia scoparia* germinate weakly in the presence of 1000 mM [7]. Despite interspecific differences in germination salt sensitivity, it appears that most halophytes retain seed viability despite the high salt concentrations to which they are subjected during dormant periods. We noted that *L. maritima* shows rapid recovery after the transfer of ungerminated seeds to distilled water. The germination rate reaches 60% in seeds previously subjected to 300 mM. Similar behaviors have been described in *Halocnumum strobilacaeum*, *Salicornia ramosissima*, *Arthrocneuum macrostachyum*, *Sarcocornia fruticosa* [8], *Allenrolfea occidentalis* [7] and *Cakile maritima* [18]. Keiffer and Ungar [19] showed that the seeds of *Salicornia europaea* and *Suaeda calceoformis* regain a recovery rate that exceeds the rate observed with the control after a two-year saline pre-treatment. All these data confirm the osmotic nature of the effects of salt on halophyte germination. The seeds of these plants would control the accumulation of Na^+ and Cl^- ions despite their abundance in the medium. This property, which ensures good protection of seeds against the deleterious effects of salt, gives halophytes an ability to grow in biotopes characterized by fluctuations in soil salinity. Indeed, the absence of germination during the period of the year characterized by high salinity without affecting the viability of the seeds is a means of preserving the seeds until the conditions become favorable at ground level, during its desalination by rainfall. In order to improve the germination of *L. maritima* seeds at 100–300 mM NaCl, we studied the effect of priming the seeds with already known germination promoting substances such as asnitrate, thiourea, proline and salicylic acid on germination in the absence or presence of a range of concentrations of NaCl—50, 100, 200 and 300 mM.

Our results showed that nitrate improves the germination of *L. maritima* seeds, especially in the presence of salt. Stimulation of germination by NO_3^- has been observed in many halophytes such as *Shoenia filifolia* [20], *Aconitum heterophyllum* [21], *Atriplex occidentalis* [22], *A. griffitii* and *Sporobolus arabicus* [23], *Paspalum vaginatum* [24] and *Crithmum maritimum* [12]. Other studies reported no effect of KNO_3 on *Crithumum maritmim*, French population [11], and *Silene mollissima* [25]. Despite the multitude of studies using KNO_3 to improve germination, the mode of action of NO_3^- on germination is poorly understood. Several hypotheses have been proposed to explain this mechanism. The first experiments by Robert and Smith [26] hypothesized that nitrate activates the phosphate pentoses pathway, resulting in stimulation of germination. Subsequent studies have shown that this compound stimulates the absorption of oxygen necessary for seed dormancy [27]. A third hypothesis postulates that NO_3^- acts as a cofactor of cytochromes [28]. Later, Hilhorst and Karssen [29] proposed a model that integrates the effect of temperature, nitrate and light on germination and dormancy. In this model, NO_3^- binds to the plasma membrane and increases the affinity of the protein (phytochrome receptor) to the active form of phytochrome (Pfr). The formation of this complex (active receptor-form of phytochrome) induces transduction signals leading to the biosynthesis of gibberellic acid strongly involved in the mobilization of reserves, on which the germination process strongly depends. These data explain the stimulation of germination that we observed in control seeds soaked in KNO_3^--enriched distilled water. Nevertheless, our results showed the beneficial effect of NO_3^- especially in the presence of salt. In addition, the results of the first experiments showed that salt subjects seeds particularly to osmotic effects. Under these conditions, we believe that the addition of NO_3^- in the imbibition solution would lead to a strong accumulation of this cation inside the cells to ensure an osmotic balance with the external environment, which is a favorable condition for imbibition and germination of seeds. Moreover, a strong positive relationship between the NO_3^- content of seeds and their germination capacity has been observed in *L. maritima* (unpublished data).

In the present study, we noticed that thiourea increases the final germination of *L. maritima* seeds if applied at salt doses of 100, 200 or 300 mM but it does not increase the rate of germination. At this level, it appears less effective than potassium nitrate. Our results are thus in agreement with those of Khan and Ungar [23] obtained on the seeds of *Sporobolus arabicus* but are in contradiction with the data acquired by the same authors on the seeds of *Atriplex griffitii* or *Aconitum heterophyllum* [21]. The stimulation of germination induced by thiourea, especially in the presence of salt, reflects an ability of this nitrogen compound to alleviate the dormancy induced by the presence of NaCl in the imbibition solution. Indeed, for a long time, the effectiveness of nitrogen compounds in alleviating primary or secondary dormancy has been demonstrated in several plant species [30] by inducing gibberellin biosynthesis pathways. According to the same authors, thiourea is involved in the alleviation of dormancy in *Avena fatua* lines, however, it remains significantly less effective than NO_3^- and its effect is strictly limited to the least dormant lines. The promoting action of thiourea on germination is related to its inhibitory effect on phenol oxidase [31], resulting in an increase in the availability of oxygen for enzymes expressing low affinity to this gas, such as the terminal oxidase of alternative respiration.

We have seen that proline, provided in the imbibition solution at the rate of 1 mM, alters the kinetics of germination and delays the germination process at all doses of the salts used. On first examination, this result is unexpected given the role of proline in osmoregulation and its involvement in the conservation of cellular structures and their protection against reactive oxygen species [32]. In this context, several authors report an improvement in the germination of halophytes such as *Zygphyllum simplex*, *Allenophlora occidentallis* and *Atriplex griffithii* [7]. Other studies report a positive relationship between the ability of seeds to germinate and their ability to accumulate proline. For example, exposure of *Kosteletzkya virginika* seeds to salt during germination is accompanied by a marked increase in proline in sprouted seeds compared to ungerminated seeds [33]. However, this proline germination promoting action has not been observed in other plant species and under various experimental conditions. Thus, Khan and Ungar [9] noted that both proline and glycine-betaine partially alleviate the primary dormancy of *Zygophyllum simplex* seeds only when germination is conducted on distilled water or at very low salinities, not exceeding 25 mM NaCl. Both nitrogen compounds do not lead to any stimulation of germination under conditions of high salinity. Similarly, Khan and Ungar [9] noted that proline and betaine do not alleviate salt-induced dormancy in both types of seeds produced by *Arthrocnemum indicum*.

The inhibition of the germination of *L. maritima* seeds by proline, regardless of the salinity of the medium, would not be due to an osmotic effect. The intake of 1 mM of proline will not significantly alter the osmotic balance between the germinating seeds and their environment. One could rather think of a toxic effect of proline. The low hydration of the seeds that would be associated with a strong endocellular accumulation of proline would bring the concentration of this compound back to toxic levels. According to some studies, the accumulation of proline is not always a trait of tolerance. For example, in rice, the accumulation of proline is associated with symptoms of toxicity and is therefore a marker of sensitivity and not tolerance [34]. In the same context, Garcia et al. [35] noted that exogenous proline intake further develops these symptoms of toxicity in rice. Proline toxicity was also observed in transgenic plants of *Arabidopsis thaliana* [32]. In order to diagnose the effects associated with a high accumulation of proline, these authors produced transgenic plants deficient in proline dehydrogenase (PDH), an enzyme involved in the degradation of proline, by means of antisense constructions. The seeds from these transgenic plants are sown in a medium with the addition of increasing doses of proline. The authors noted an increased sensitivity of transgenic plants at the germination stage and post-germination growth. The authors have highlighted in particular a halt in the development of seedlings associated with cell mortality. They thus concluded that inhibition of the catabolic pathways of proline leads to toxic proline contents, up to 1068 μg of proline per g of fresh matter. In light of these data, we believe that a strong accumulation of proline in the germinating seeds of *L. maritima*, probably associated with limited activity of the amino acid's breakdown enzymes, would subject the seeds to toxic effects.

Salicylic acid was found to have no significant effect on the germination of *L. maritima* seeds at 0 mM NaCl. However, it seems to amplify the depressive effect of salt at higher salt concentrations. Our results are consistent with those obtained by Borsani et al. [36] on transgenic plants of *Arabidopsis thaliana*. By comparing the response of a wild lineage of *A. thaliana* to that of a transgenic line deficient in salicylic acid (integrating a trangen encoding salicylate hydrolase), these authors observed a particular sensitivity of the wild line subjected during germination or post-germination growth to 100 mM NaCl or 270 mM of mannitol. They conclude that salicylic acid enhances the effects of salinity and osmotic stress by increasing the genesis of reactive oxygen species during photosynthesis and germination in *A. thaliana*. However, the effect of salicylic acid seems to depend on its content in the tissues. Thus, several studies report a beneficial effect of salicylic acid on the response of plants to abiotic or biotic constraints. For example, incubation of *Raphanus sativus* seeds in a salicylic acid solution increases the optimal temperature range for germination [37]. The pre-treatment of maize seeds with salicylic acid induces antioxidant enzymes, resulting in an increase in its resistance to cold [38]. Salicylic acid improved the germination of *Leynus chinensis* under salt-alkali stress through an osmotic priming role [38].

5. Conclusions

Overall, it can be concluded that the germination of *L. maritima* seeds is improved by potassium nitrate in the presence or absence of salt, while thiourea increases the germination rate without affecting the germination rate. Salicylic acid amplifies the effect of salt, this effect is all the more important as the salinity of the imbibition solution increases. Whereas, proline delays germination without stopping it completely even after 46 days of treatment. These findings indicate that the application of KNO_3 and thiourea may be used to improve seed germination of *L. maritima*, which is of great interest for cultivating this plant for landscaping purposes of saline soils.

Author Contributions: K.B.H.: Conceptualization, data curation, writing—original draft preparation, writing—review and editing, figures, supervision, project coordination. I.Z., A.D. and S.Y.: Data curation, writing—original draft preparation, figures. All authors have read and agreed to the published version of the manuscript.

Funding: Financial support of the project "HaloFarMs" by PRIMA, a program supported by the European Union, and by the Ministry of Higher Education and Scientific Research (MHES), Tunisia.

Institutional Review Board Statement: Not applicable.

Informed Consent Statement: Not applicable.

Data Availability Statement: Not applicable.

Conflicts of Interest: The authors declare no conflict of interest.

References

1. Wang, K.; Wu, Y.-H.; Tian, X.-Q.; Bai, Z.-Y.; Liang, Q.-Y.; Liu, Q.-L.; Pan, Y.-Z.; Zhang, L. Overexpression of DgWRKY4 enhances salt tolerance in chrysanthemum seedlings. *Front. Plant Sci.* **2017**, *8*, 1592. [CrossRef] [PubMed]
2. Zhang, B.; Xie, L.; Sun, T.; Ding, B.; Li, Y.; Zhang, Y. *Chrysanthemum morifolium* aquaporin genes CmPIP1 and CmPIP2 are involved in tolerance to salt stress. *Sci. Hortic.* **2019**, *256*, 108627. [CrossRef]
3. García-Caparrós, P.; Llanderal, A.; Lao, M.T. Halophytes as an Option for the Restoration of Degraded Areas and Landscaping. In *Handbook of Halophytes*; Grigore, M.-N., Ed.; Springer Nature: Cham, Switzerland, 2020. [CrossRef]
4. Santos, J.; Al-Azzawi, M.; Aronson, J.; Flowers, T.J. eHALOPH a database of salt tolerant plants: Helping put halophytes to work. *Plant Cell Physiol.* **2016**, *57*, e10. [CrossRef] [PubMed]
5. Cassaniti, C.; Romano, D. The use of halophytes for Mediterranean landscaping. *Eur. J. Plant Sci. Biotechnol.* **2011**, *5*, 58–63.
6. Gul, B.; Ansari, R.; Flowers, T.J.; Khan, M.A. Germination strategies of halophyte seeds under salinity. *Environ. Exp. Bot.* **2013**, *92*, 4–18. [CrossRef]
7. Gul, B.; Hameed, A.; Weber, D.J.; Khan, M.A. Assessing Seed Germination Responses of Great Basin Halophytes to Various Exogenous Chemical Treatments Under Saline Conditions. In *Sabkha Ecosystems*; Khan, M., Boër, B., Őzturk, M., Clüsener-Godt, M., Gul, B., Breckle, S.W., Eds.; Springer: Cham, Switzerland, 2016; Volume 48. [CrossRef]
8. Pujol, J.A.; José, F.C.; Luis, R. Recovery germination from different osmotic conditions by four halophytes from South-eastern Spain. *Ann. Bot.* **2000**, *85*, 279–286. [CrossRef]
9. Khan, M.A.; Ungar, I.A. Action of compatible osmotica and growth regulators in alleviating the effect of salinity on the germination of dimorphic seeds of *Arthrocneumum indicum*. *Int. J. Plant Sci.* **1998**, *159*, 313–317. [CrossRef]
10. Meot-Duros, L.; Magné, C. Effect of salinity and chemical factors on seed germination in the halophyte *Crithmum maritimum* L. *Plant Soil.* **2008**, *313*, 83–87. [CrossRef]
11. Atia, A.; Debez, A.; Barhoumi, Z.; Smaoui, A.; Abdelly, C. ABA, GA 3, and nitrate may control seed germination of *Crithmum maritimum* (Apiaceae) under saline conditions. *Comptes Rendus Biol.* **2009**, *332*, 704–710. [CrossRef]
12. Atia, A.; Barhoum, Z.; Rabhi, M.; Abdely, C.; Smaoui, A. Environmental eco-physiology and economical potential of the halophyte *Crithmum maritimum* L. (Apiaceae). *J. Med. Plants Res.* **2011**, *5*, 3564–3571.
13. Bouzidi, A.; Krouma, A.; Chaieb, M. Chemical seed priming alleviates salinity stress and improves *Sulla carnosa* germination in the saline depression of Tunisia. *Plant Direct* **2021**, *5*, e357. [CrossRef] [PubMed]
14. Picó, F.X.; Retana, J. The flowering pattern of the perennial herb *Lobularia maritima*: An unusual case in the Mediterranean basin. *Acta Oecologica* **2001**, *22*, 209–217. [CrossRef]
15. Brennan, E.B. Agronomic aspects of strip intercropping lettuce with alyssum for biological control of aphids. *Biol. Control.* **2013**, *65*, 302–311. [CrossRef]
16. Tiwari, S.; Saville, D.J.; Wratten, S.D. Preferences of the wheat bug (*Nysius huttoni*) for particular growth stages of the potential trap crop, alyssum (*Lobularia maritima*). *N. Z. Plant Prot.* **2019**, *72*, 237–244. [CrossRef]

17. Gulzar, S.; Khan, M.A. Effects of salinity and temperature on the germination of *Urochondra stulosa* (trin.) C.E. Hubbard. *Seed Sci. Technol.* **2001**, *29*, 21–29.
18. Debez, A.; Ben Hamed, K.; Grignon, C.; Abdelly, C. Salinity effects on germination, growth, and seed production of the halophyte *Cakile maritima*. *Plant Soil* **2004**, *262*, 179–189. [CrossRef]
19. Keiffer, C.H.; Ungar, I.A. Germination responses of halophytes seeds exposed to prolonged hypersaline conditions. In *Biology of Salt Tolerant Plants*; Khan, M., Ungar, I.A., Eds.; Pakistan University of Karachi: Karachi, Pakistan, 1996; pp. 43–50.
20. Plumer, J.A.; Rogers, A.D.; Turner, D.W.; Bell, D.T. Light, nitrogen compounds, smoke and GA 3 break dormancy and enhance germination in the Australian everlasting daisy, *Shoenia filifolia* Sbsp. *Subulifolia*. *Seed Sci. Technol.* **2001**, *29*, 321–330.
21. Pandey, H.; Nandi, S.K.; Palni, L.M.S. Chemical stimulation of seed germination in *Aconitum heterophyllum* Wall. & A. Balfourii Stapf: Important Himalayan species of medicinal value. *Seed Sci. Technol.* **2000**, *28*, 39–48.
22. Gul, B.; Weber, D.J. Effect of dormancy relieving compounds on the seed germination of non dormant *Allenrolfea occidentalis* under salinity stress. *Ann. Bot.* **1998**, *82*, 555–560. [CrossRef]
23. Khan, M.A.; Ungar, I.A. Alleviation of innate and salinity–induced dormancy in *Atriplex griffithii* Moq. Var. Stocksii Boiss. *Seeds Sci. Technol.* **2000**, *28*, 29–37.
24. Shim, S.I.; Moon, J.-C.; Jang, C.S.; Raymer, P.; Kim, W. Effect of potassium nitrate priming on seed germination of seashore paspalum. *HortScience* **2008**, *43*, 2259–2262. [CrossRef]
25. Murru, V.; Santo, A.; Piaza, C.; Hugit, L.; Bacchetta, G. Seed germination, salt stress tolerance, and the effect of nitrate on three Tyrrnian coastal species of the *Silene mollissima* aggregate (Caryophyllaceae). *Botany* **2015**, *93*, 881–892. [CrossRef]
26. Hilton, J.R.; Thomas, J.A. Regulation of pregerminative rates of respiration in seeds of various seed species by potassium nitrate. *J. Exp. Bot.* **1996**, *37*, 1516–1524. [CrossRef]
27. Hilhorst, H.W.M.; Karseen, C.M. Dual Effect of light on the gibberelin and nitrate stimulated seed germination of *Sisymbryom officinale* and *Arabidopsis thaliana*. *Plant Physiol.* **1988**, *86*, 591–597. [CrossRef]
28. Hilhorst, H.W.M.; Karssen, M. Effect of chemical environment in seed germination. In *The Ecology of Regeneration in Plant Communities*, 2nd ed.; Fenner, M., Ed.; CAB International Wallingford: Wallingford, UK, 2000; pp. 293–311.
29. Adkins, S.; Graham, M.; Simpson, G.; Naylor, J.M. The physiological basis of seed dormancy in *Avena fatua*. *Physiolgia Plant.* **1984**, *60*, 234–238. [CrossRef]
30. Roberts, E.H.; Smith, R.D. Dormancy and the pentose phosphate pathway. In *The Physiology and Biochemistry of Seed Dormancy and Germination*; Khan, A.A., Ed.; Elsevier: Amsterdam, The Netherlands; North-Holland Biomedical Press: New York, NY, USA, 1977; pp. 385–411.
31. Mani, S.; Van de cotte, B.; Van Montagu, N.; Verbruggen, N. Altered Levels of Proline Dehydrogenase Cause Hypersensivity to Proline and Its analogue in Arabidopsis. *Plant Physiol.* **2002**, *128*, 73–83. [CrossRef] [PubMed]
32. Poljakoff-Mayber, A.; Somers, G.F.; Werker, E.; Gallagher, J.L. Seeds germination of *Kosteletzka virginica* germination and salt tolerance. *Am. J. Bot.* **1994**, *81*, 54–59. [CrossRef]
33. Lutts, S.; Kinet, J.M.; Bouharment, J. NaCl-induced senescence in leavesof rice (*Oryza sativa*) cultivar differing in salinity resistance. *Ann. Bot.* **1996**, *78*, 389–398. [CrossRef]
34. Garcia, A.B.; De Almeider, E.J.; Iyer, S.; Gerats, T.; Van Mantagu, M.; Caplan, A.B. Effects of osmoprotectants upon NaCl stress in rice. *Plant Physiol.* **1997**, *115*, 159–169. [CrossRef]
35. Borsani OValpuesta, A.; Miguel, A.B. Evidence for a Role of Salicylic Acid in the Oxidative Damage Generated by and Osmotic Stress in *Arabidopsis* Seedlings. *Plant Physiol.* **2001**, *126*, 1024–1030. [CrossRef]
36. Takaki, M.; Rosim, R.E. Aspirin increases tolerance to high temperature in seeds of *Raphanus sativus* L. cVAR Early Scarlet Globe. *Seed Sci. Technol.* **2000**, *28*, 179–183.
37. Janda, T.; Sazalai, G.; Tari, I.; Pàldi, E. Hydroponic treatement with salisalic acid decreases the effects of chiling injury in maize.(*Zea mays* L.) Plants. *Planta* **1999**, *208*, 175–180. [CrossRef]
38. Hongua, C.; Leynan, T.; Junmei, S.; Xiaou, H.; Xianguo, C. Exogenous salicylic acid reveals an osmotic regulatory role in priming the seed germination of *Leymus chinensis* under salt-alkali stress. *Environ. Exp. Bot.* **2021**, *18*, 104498. [CrossRef]

seeds

MDPI

Article

Stimulating Role of Calcium and Cyclic GMP in Mediating the Effect of Magnetopriming for Alleviation of Salt Stress in Soybean Seedlings

Sunita Kataria [1,*], Shruti Shukla [1], Kanchan Jumrani [2], Meeta Jain [1,*] and Rekha Gadre [1]

1 School of Biochemistry, Devi Ahilya Vishwavidyalaya, Khandwa Road, Indore 452001, MP, India
2 Division of Plant Physiology, Indian Institute of Soybean Research, Khandwa Road, Indore 452001, MP, India
* Correspondence: sunita_kataria@yahoo.com (S.K.); meetajainind@yahoo.com (M.J.)

Abstract: This current study examined the role of calcium (Ca) and Cyclic GMP (cGMP) in mitigating the adverse effect of salt stress through magnetopriming of soybean cultivar JS-335 seeds with a static magnetic field (SMF, 200 mT for 1 h). The salt stress (50 mMNaCl) extensively reduced the early seedling growth (64%), vigour Index-I (71%), vigour Index-II (39%), total amylase (59%), protease (63%), and nitrate reductase (NR, 19%) activities in un-primed soybean seedlings. However, magnetopriming and Ca treatment enhanced all of these measured parameters along with remarkable increase in reactive oxygen species (ROS) and nitric oxide (NO) content. The exogenous application of Ca^{2+}, cGMP and ROS regulators such as nifedipine (Ca^{2+} channel blocker), EGTA, ethylene glycol-β-amino ethyl ether tetra acetic acid (Ca^{2+}chelators), genistein (cGMP blocker), and dimethyl thiourea (DMTU, H_2O_2 inhibitor) negatively affects the SMF-induced seedling length, seedling vigour, ROS, NO, and enzyme activities such as protease, total amylase, and NR in soybean seedlings. Results presented by using specific various biochemical inhibitors of Ca, cGMP, or ROS signalling in vivo indicated that Ca and cGMP are also involved with ROS and NO in the signal transduction of magnetic field enthused soybean seed germination and seedling growth under salt stress.

Keywords: calcium; cGMP; magnetic field; reactive oxygen species; salt stress

Citation: Kataria, S.; Shukla, S.; Jumrani, K.; Jain, M.; Gadre, R. Stimulating Role of Calcium and Cyclic GMP in Mediating the Effect of Magnetopriming for Alleviation of Salt Stress in Soybean Seedlings. *Seeds* **2023**, *2*, 232–245. https://doi.org/10.3390/seeds2020018

Academic Editors: José Antonio Hernández Cortés, Gregorio Barba-Espín and Pedro Diaz-Vivancos

Received: 22 April 2023
Revised: 10 May 2023
Accepted: 11 May 2023
Published: 16 May 2023

1. Introduction

During the last decade the world has experienced an unforeseen change in climate. Salinity is the utmost brutal environmental stressor limiting crop productivity as most of the crops are sensitive to salinity because of high concentrations of salts in the soil [1–3]. Salinization is increasing worldwide and there is decline in the average yields for most crops. Seed germination and emergence are the most vulnerable stages of the plant growth cycle under stress conditions. Various researchers have reported that salt stress may affect germination of seed by generating an osmotic stress thus averting the seed from uptake of water or through the toxic effects of sodium and chloride ions [4–6]. Additionally, salt stress can accelerate various stresses such as oxidative stress, ionic stress, osmotic stress, and hormonal imbalances which affects cell functions, leading to cell damage which ultimately slowdown plant growth [4,5,7].

Plants are able to recognize stimuli and recourse in response to different abiotic environmental stresses by activating defense mechanisms. Various pre-sowing treatments such as magnetopriming which is a non-invasive technique in which dry seeds are treated with a static magnetic field (SMF), improved the seed vigour and germination under various abiotic stress conditions [8–11]. The positive impact of SMF and enrichment of seed germination due to magnetic field (MF) exposure has been observed by various researchers in different crops such as peach (*Prunus persica*) [12], chickpea (*Cicer arietinum*) [13], cucumber (*Cucumis sativus*) [14], lettuce (*Lactuca sativa*) [15], corn (*Zea mays*) [16], tomato (*Solanum lycopersicum*) [17], radish (*Raphanus sativus*) [18], soybean (*Glycine max*), and maize (*Zea mays*) [19]. The effect of magnetic bio-stimulation of seeds under salt stress using SMF was reported in chickpea, soybean, and maize [19–22]. The SMFpre-treatment enhances the tolerance against salt stress by increase in the activity of certain enzymes such as amylase, protease, dehydrogenase, and nitrate reductase [19–22]. Further, Kataria et al. [22] also suggested that magnetoprimed seeds maintain a balance of the plant hormones abscisic acid (ABA), gibberellic acid (GA), and indoleacetic acid (IAA) through the signalling molecule nitric oxide (NO), which helps to counterbalance the negative effects of salinity on seed germination and growth of soybean.

Soybean is the most widely cultivated seed legume and provides food, edible oil, protein concentrate for livestock feeding, constituent in formulated diet of fish and poultry, and a variety of industrial products [23]. Soybean is adapted to a wide range of climate and soil conditions; it is sensitive to various abiotic stresses such as drought, high temperature, metal toxicity, and salinity [8]. Among various abiotic stress, soil salinity is one of the major threat to soybean productivity [1]. Salt stress is answerable for the delayed or reduced germination and vigour of soybean seedlings [1,21,22]. Salinity stress negatively affects the seed germination, plant growth, photosynthesis, biomass accumulation, and ultimately quality and yield of soybean [1,21,22].

One of the intriguing components of magnetic field treatment is that they seemed to upgrade resistance to abiotic [24] or biotic [25] stresses as a consequence of the cell reinforcement reaction actuation. There are reports on the positive effect of magnetopriming on seed germination, early seedling growth, photosynthesis, PS II efficiency, plant growth, and yield on crops such as barley, soybean, wheat, maize, and mungbean under various abiotic stresses including drought, cadmium, salt, and UV-B stress [8,19,21,26–29]. An alleviation strategy, for example magnetopriming, used to safeguard the soybean seedlings from the adverse effects of salinity has been previously identified [21,22]. Further, the role of ROS and NO signalling was reported in the soybean seeds under salt and UV-B stress [21,30]. In addition to ROS and NO, it has also been observed that calcium ions (Ca^{2+}) and cyclic Guanosine Monophosphates (cGMPs) are known to play an essential role in the process of signal transduction in plants along with a role in the growth and development [31] under biotic and abiotic stress.

Calcium acts as a universal signal molecule to play pivotal roles in plant growth and development, including cell wall formation [32], osmotic regulation [33], cell division [34], and resistance to biotic and abiotic stresses [35–37]. It has been observed that calcium is an important element for salt tolerance and deliberates a protective role in plants growing under sodic soils. Under stressful conditions, the formation of hydrogen peroxide (H_2O_2) and NO increases at the same time and acts together to achieve vital cellular functions [38–40]. Furthermore, in response to abiotic stresses, H_2O_2 and NO molecules participate in cross-talk with Ca^{2+} to form a complex signalling network [38,40]. It has been reported that cross-talk between H_2O_2 and NO embolden plant tolerance to salinity stress [40,41]. Therefore, our investigation intended to evaluate the role of Ca^{2+} and cGMP along with ROS in the salt stress management in magnetoprimed soybean seeds by considering the seed germination and early growth attributes of seedlings. To accomplish this we used disparate Ca^{2+} and cGMP regulators such as calcium chloride ($CaCl_2$), nifedipine (Ca^{2+} channel blocker), EGTA, i.e., ethylene glycol-β-amino ethyl ether tetra acetic acid (Ca^{2+} chelators), genistein (cGMP blocker), and dimethyl thiourea (DMTU, H_2O_2 inhibitor).

2. Materials and Methods

2.1. Plant Material

Soybean seeds (*Glycine max* [L.] cultivar JS-335) (semi-determinate; having purple flowers, yellow seed coat, and days to maturity is about 100 days) were used for the present study as this variety of soybean seeds are most sensitive to salt stress and provides the best result in response to magnetic field pre-treatment under salt stress [19,21,24,42]. The plant material was obtained from ICAR (Indian Institute of Soybean Research, Khandwa Road, Indore, India).

2.2. Treatment under Magnetic Field

Pre-treatment of dry soybean seeds cultivar JS-335 under MF was performed using a fabricated electromagnetic field generator ("AETec" Academy of Embedded Technology, Delhi, India). The exposure was carried out under SMF strength of 200 mT for a time period of about 1 h under ambient conditions of temperature at 25 ± 5 °C based upon previous studies [19]. During the experiments, un-primed seeds were kept away from the impact of MF. All the experiments were accomplished parallel to the control (un-primed seeds) in the laboratory of Biochemistry department, Devi Ahilya University, Indore (22.7196° N, 75.8577° E), Madhya Pradesh, India.

2.3. Screening of Concentration of Different Modulators of cGMP and $CaCl_2$

In order to select the appropriate concentration of modulators to be used during the experiments, screening of these modulators was performed prior to germination of seeds. The modulators of cGMP, $CaCl_2$ and ROS used for the experiment are $CaCl_2$, EGTA, genistein, nifedipine, cGMP analogue, and DMTU. Based on previous research conducted on various concentrations of Ca, cGMP and ROS modulators [43–45] were taken as follows:- $CaCl_2$ (10 mM, 1.0 mM, 0.1 mM), Nef (10 mM, 1.0 mM, 0.1 mM), EGTA (10 mM, 1.0 mM, 0.1 mM), Gen (0.05 mM, 0.1 mM, 1 mM), and DMTU (0.1 mM, 0.05 mM, 1.0 mM). All of these concentrations were used for early seedling growth determination to decide the final concentration of each modulator. The seeds were germinated for five days in appropriate conditions of temperature about 25 ± 1°C. After five days (imbibitions for 120 h), different growth parameters such as seedling length, seedling dry and fresh weight, vigour index-I and II were noted and on the basis of results, the final concentrations of the modulators were determined for further experiments as 1mM $CaCl_2$, 1.0 mM Nef, 0.1 mM EGTA, 0.05 mM Gen, and 0.1 mM DMTU.

Germination of the primed and un-primed seeds was completed by surface sterilizing the seeds in 0.01% $HgCl_2$ for 2 min and washing these3–4 times with distilled water. These sterilized seeds were kept in Petri-plates (15 cm diameter) lined by Whatman filter paper No.1 containing 10 mL of distilled water. For inducing salt stress (saline condition), seeds were placed in Petri-plates (15 cm diameter) containing 10 mL of 50 mM NaCl solution. The primed and un-primed seeds in the absence and presence of NaCl were grown along with the modulators of Ca, cGMP and ROS such as calcium chloride ($CaCl_2$), genistein (Gen), nifedipine(Nef), Ethylene Glycol bis-(β-amino ethyl ether) tetra acetic acid (EGTA) and dimethylthiourea (DMTU).

2.4. Early Seedling Growth Characteristics of Seedlings

Primed and un-primed seeds were arbitrarily selected from each treatment in triplicates (n = 3) to estimate the early seedling growth parameters such as length of root, shoot and seedling, fresh and dry weight of the soybean seedlings after five days of imbibitions in different modulators of $CaCl_2$, cGMP and ROS. Seedlings were dried in an oven at 70–80 °C and dry weight was taken by the weighing balance. The vigour index of seedlings was estimated as described by Abdul-Baki and Anderson [46].

$$\text{Vigour index I} = \text{Germination } \% \times \text{Seedling length (Root + Shoot) cm}$$

$$\text{Vigour index II} = \text{Germination } \% \times \text{Seedling dry weight (Root + Shoot) g}$$

2.5. Biochemical Analysis of Seeds

All of the biochemical analysis was carried out in the randomly selected seedlings from each triplicate after five days (120 h seed imbibitions) under different treatments of modulators used under both non-saline and saline conditions.

2.5.1. Total Amylase Activity

The total amylase activity was assayed by the Sawhney et al. [47] method. Seedlings (100 mg) were homogenized in 5 mL of 80% chilled acetone, then centrifuged at 15,000 rpm for 10 min at 4 °C. The pellet was dissolved in 10 mL of 0.2 M phosphate buffer (pH 6.4) and again centrifuged at 15,000 rpm for 20 min at 4 °C. The supernatant thus obtained was mixed with 2.0 mL phosphate buffer (pH 6.4), 1.0 mL starch (1%), and incubated for 30 min at room temperature. After incubation, 0.1 N HCl and 0.1 N potassium iodide was added to the mixture and absorbance was recorded at 660 nm. The enzyme activity was expressed as mg starch hydrolysed mg^{-1} protein h^{-1}.

2.5.2. Protease Activity

Protease activity was determined using the Kunitz method [48]. Seedlings (1.0 g) were crushed in 0.2 M phosphate buffer (pH 7.4) and centrifuged at 13,800 rpm at 4 °C for 30 min. Supernatant (0.5 mL) mixed with 0.5 mL of 1% casein prepared in 0.2 M carbonate buffer and was incubated for 10 min at 37 °C and then 1.0 mL of 10% TCA was added to stop the reaction. This mixture was then centrifuged at 12,000 rpm for10 min at 4 °C. To the centrifuged solution, 2.5 mL of carbonate buffer along with 0.5 mL of Folin reagent was added and the development of the orange colour was observed and incubated for 30 min at room temperature, then the protein content was measured at 660 nm. The activity was expressed as mg protein hydrolysed g^{-1} fresh weight.

2.5.3. Nitrate Reductase Activity

The assay was carried out in seedling tissues based upon the Jaworski method [49]. The chopped seedlings (250 mg) were homogenized in 0.1 M phosphate buffer containing 25% isopropanol and 0.2 M potassium nitrate. The mixture was incubated for 2 h at 30 °C. Next, 0.1% NED and 1.0% sulphanilamide was added to the solution post-incubation, and absorbance was recorded based upon the amount of nitrite formed at 540 nm. The enzyme activity was noted as n mole of NO_2 g^{-1} FW h^{-1}.

2.5.4. Estimation of Nitric Oxide

NO estimation was performed by measuring the amount of nitrite formed using the method of Zhou et al. [50] with minor modifications. The seedling tissues (250 mg) of germinated seedlings under different treatments were homogenized in 2.0 mL of 50 mM acetate buffer (50 mM sodium acetate, 50 mM acetic acid, and 4% zinc acetate) (pH 7.6) and centrifuged at 12,000 rpm for 15 min at 4 °C, then the supernatant was saved. The pellet obtained was washed with 0.5 mL acetate buffer and then centrifuged again. Both the supernatants thus obtained were mixed together and filtered using Whatmann filter paper after adding 100 mg of charcoal. The filtrate (1 mL) was mixed with Greiss reagent (1% sulphanilamide and 0.1% NED prepared in 5% H_2PO_4) in the ratio of 1:1 and was incubated for 30 min at room temperature, and then the absorbance was recorded at 540 nm. The NO content was expressed as n moles g^{-1} FW. The standard curve was prepared using sodium nitrate.

2.5.5. Estimation of ROS

Estimation of Superoxide Radical

Superoxide radical estimation was achieved based upon the method given by Chiatanya and Naithani [51] through its ability to reduce Nitrobluetetrazolium chloride (NBT). The seedlings (100 mg) were homogenized in 2.0 mL of pre-chilled 0.2 M phosphate buffer (pH 7.2) containing DTDC (Dithio Carbonic Acid) to inhibit SOD activity. The mixture was centrifuged at 10,000 rpm for 10 min at 4 °C. The supernatant obtained was used for measuring the enzyme activity by recording the absorbance at 540 nm. The activity was measured using extinction coefficient 12.8 mM^{-1} cm^{-1} and represented as µmole of superoxide g^{-1} FW.

Estimation of Hydrogen Peroxide

Hydrogen peroxide content was estimated using Mukherjee and Choudhary [52] by measuring titanium hydro-peroxide complex. The seedlings (500 mg) were homogenized in chilled acetone and filtered by Whatmann No. 1 filter paper. To the homogenate obtained, 2.0 mL titanium reagent (titanium oxide and potassium sulphate processed in concentrated sulphuric acid) and 2.5 mL of ammonium hydroxide solution was added to the filtrate for precipitation of the titanium hydro-peroxide complex. The complex was centrifuged at 13,800 rpm for 15 min at 4 °C. The pellet that was obtained was re-centrifuged after adding 2 M sulphuric acid. After centrifugation, absorbance of the supernatant obtained was noted at 415 nm and the activity was described as µmoles of H_2O_2 g^{-1} FW.

2.6. Statistical Analysis

These data are expressed as mean ± S.E. (n = 3) and were analysedusingthe analysis of variance (ANOVA) followed by the post hoc Newman–Keuls Multiple Comparison Test. A ### $p < 0.001$, ## $p < 0.01$, and # $p < 0.05$ indicates the significant difference amongst the un-primed seeds grown in the control with the un-primed seeds grown in different modulators under non-saline or saline conditions and *** $p < 0.001$, **$p < 0.01$, and * $p < 0.05$ indicates the significant difference amongst the magnetoprimed seeds grown in control with the magnetoprimed seeds grown in different modulators under non-saline or saline conditions by using Prism 4 software for Windows, GrafPad Software, San Diego, CA, USA.

3. Results and Discussion

Calcium is an essential mineral nutrient for plant growth and development. It plays a pivotal role in maintaining the structural and functional integrity of plant membranes, stabilizing cell walls, controlling ion transport, and regulating ion-exchange behaviour and cell wall enzyme activities [50,51]. We found that 50 mM NaCl remarkably decreased the seedling length (64%), vigour Index-I (71%),vigour Index-II (39%), total amylase (59%), protease (63%), and NR (19%), and ROS content were increased in the seedlings of un-primed seeds while magnetopriming significantly enhanced all of these parameters in non-saline as well as saline conditions (Figures 1–3). To identify the role of Ca^{2+} and cGMP in magnetopriming induced signalling at the alleviation of salt stress in soybeans, we used disparate Ca^{2+} and cGMP regulators such as Nef, nifedipine, 1 mM (Ca^{2+} channel blocker), EGTA, ethylene glycol-β-amino ethyl ether tetra acetic acid-0.1 mM (Ca^{2+}chelators), $CaCl_2$ 1 mM (Ca^{2+} analogue), genistein 0.05 mM (cGMP blocker), and dimethyl thiourea 0.1 mM (DMTU). We found that when seedlings were grown in calcium ($CaCl_2$) the seedling length, and vigour Index-I and II were significantly enhanced in SMF-primed seeds in comparison to un-primed seeds under both the non-saline (Figure 1A–C) and saline conditions (Figure 1D–F). However, the extent of promotion was more in seedlings from SMF-primed seeds of Ca ($CaCl_2$ treatment) grown under saline conditions; it showed 24% promotion in vigour Index-I (Figure 1E) and 20% in vigour Index-II as compared to their SMF-primed seedlings grown in controls (Figure 1F). After EGTA, Nef, Gen, and DMTU treatments, all the measured early seedling growth parameters were reduced to a greater extent in SMF-primed seeds in comparison to un-primed seeds under both the conditions (Figure 1A–F).

When compared to un-primed seeds, the SMF treatment caused an increase in $O_2^{\bullet-}$, H_2O_2, and NO content in soybean seedlings (Figure 2A–C) under non-saline and saline conditions (Figure 2D–F). After EGTA, Nef, Gen, and DMTU treatment the ROS and NO content was reduced to a greater extent in SMF-primed seeds in comparison to un-primed seeds in both the conditions (Figure 2A–F). In the absence of salt stress, the SMF-induced $O_2^{\bullet-}$content was reduced by DMTU (68%), Nef (57%), Gen (46%), and EGTA (40%) (Figure 2A). Whereas, under salt stress, the $O_2^{\bullet-}$ level was reduced by 66, 61, 54, and 46% by DMTU, Gen, EGTA, and Nef, respectively, in SMF-primed seedlings (Figure 2D). Amongst Ca^{2+} blockers, Nef caused the maximal reduction with 50% and 62.5% in SMF-induced H_2O_2 production in non-saline and saline conditions, respectively (Figure 2B,E). cGMP blocker also triggered significant reduction in SMF-induced H_2O_2 production, i.e.,

57% and 71%, respectively, in non-saline and saline conditions (Figure 2B,E). The changes in $O_2^{\bullet -}$, H_2O_2 and NO levels was not significant in the seedlings of un-primed seeds upon inhibitors treatments except for DMTU (Figure 2A–F).

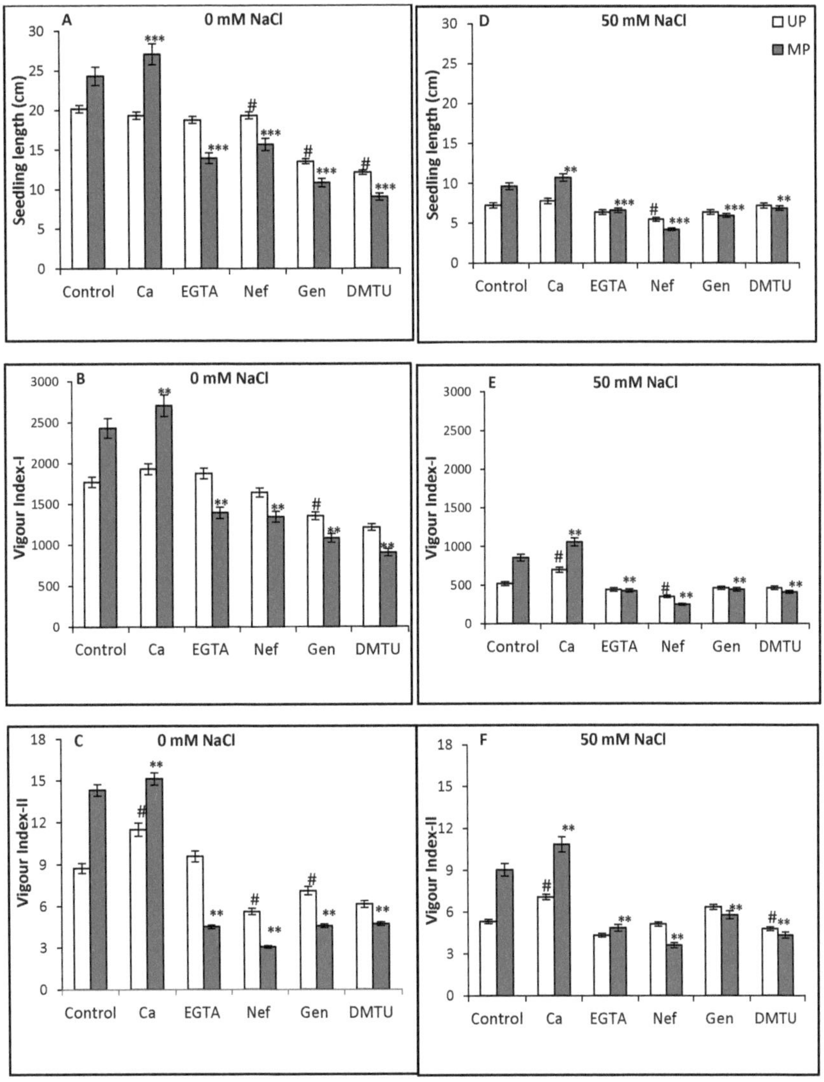

Figure 1. Effects of different modulators of calcium, cGMP, and ROS on SMF pre-treatment (200 mT for 1 h) induced seedling length (**A**,**D**), vigour Index-I (**B**,**E**), and vigour Index-II (**C**,**F**) of soybean seedlings, respectively, in non-saline and saline conditions. These data are expressed as mean ± SE ($n = 3$). # $p < 0.05$ indicates the significant difference amongst the UP seeds grown in the control with the UP seeds grown in different modulators under non-saline or saline conditions and *** $p < 0.001$, ** $p < 0.01$ indicates the significant difference amongst the MP seeds grown in control with the MP grown in different modulators under non-saline or saline conditions. UP, un-primed seeds and MP, magnetoprimed seeds with SMF. Ca, calcium chloride ($CaCl_2$), 1 mM (Ca^{2+} analogue); EGTA, ethylene glycol-β-amino ethyl ether tetra acetic acid, 0.1 mM; Nef, nifedipine,1 mM (Ca^{2+} channel blocker); Gen, genistein, 0.05 mM (cGMP blocker),and dimethyl thiourea, 0.1 mM (DMTU, H_2O_2 inhibitor).

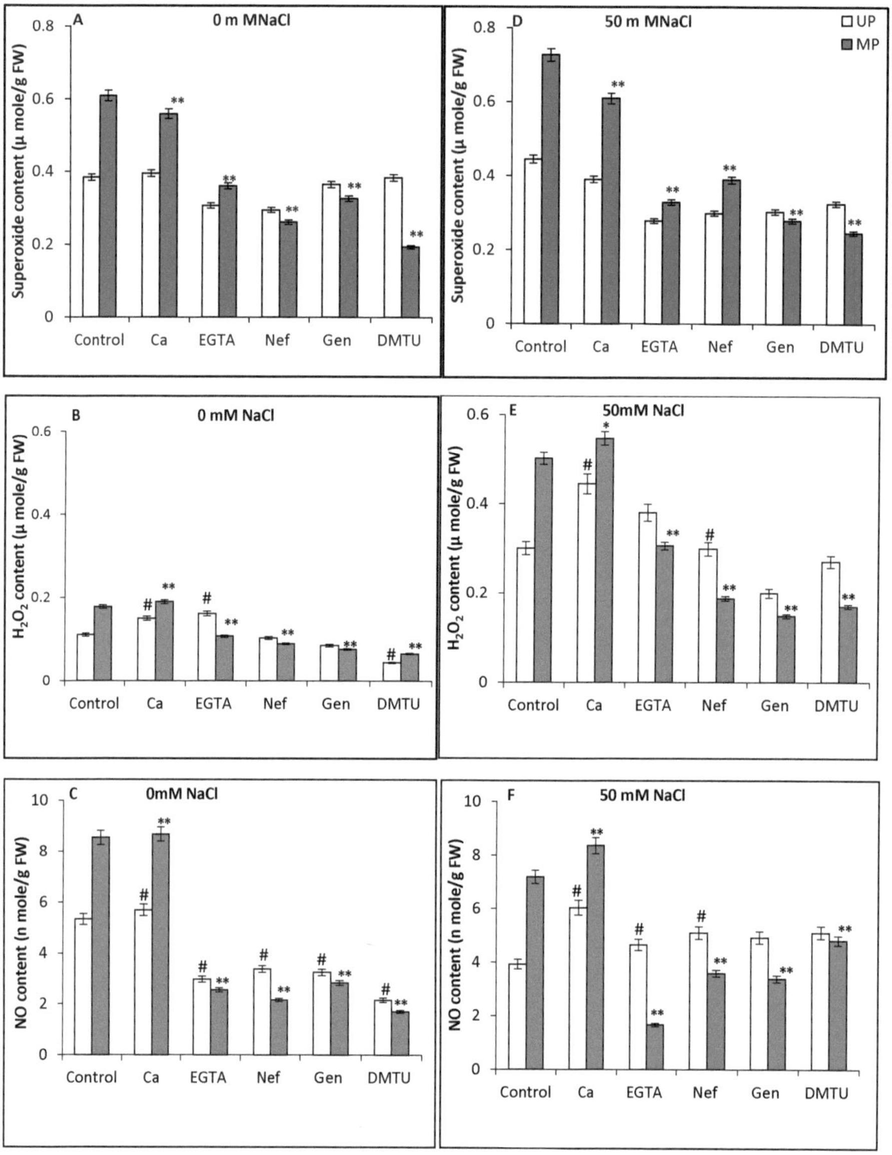

Figure 2. Effects of different modulators of calcium and cGMP on SMF pre-treatment (200 mT for 1 h) induced superoxide radical(**A**,**D**), hydrogen peroxide (**B**,**E**) and nitric oxide (**C**,**F**) content in soybean seedlings, respectively, in non-saline and saline conditions. These data are expressed as mean + SE (n = 3). # p< 0.05 indicates the significant difference amongst the UP seeds grown in control with the UP seeds grown in different modulators under non-saline or saline conditions and ** p < 0.01, * p < 0.05 indicates the significant difference amongst the MP seeds grown in control with the MP grown in different modulators under non-saline or saline conditions. UP, un-primed seeds and MP, magnetoprimed seeds. Ca, calcium chloride ($CaCl_2$), 1 mM (Ca^{2+} analogue); EGTA, ethylene glycol-β-amino ethyl ether tetra acetic acid, 0.1 mM; Nef, nifedipine, 1 mM (Ca^{2+} channel blocker); Gen, genistein, 0.05 mM (cGMP blocker), and dimethyl thiourea, 0.1 mM (DMTU H_2O_2 inhibitor).

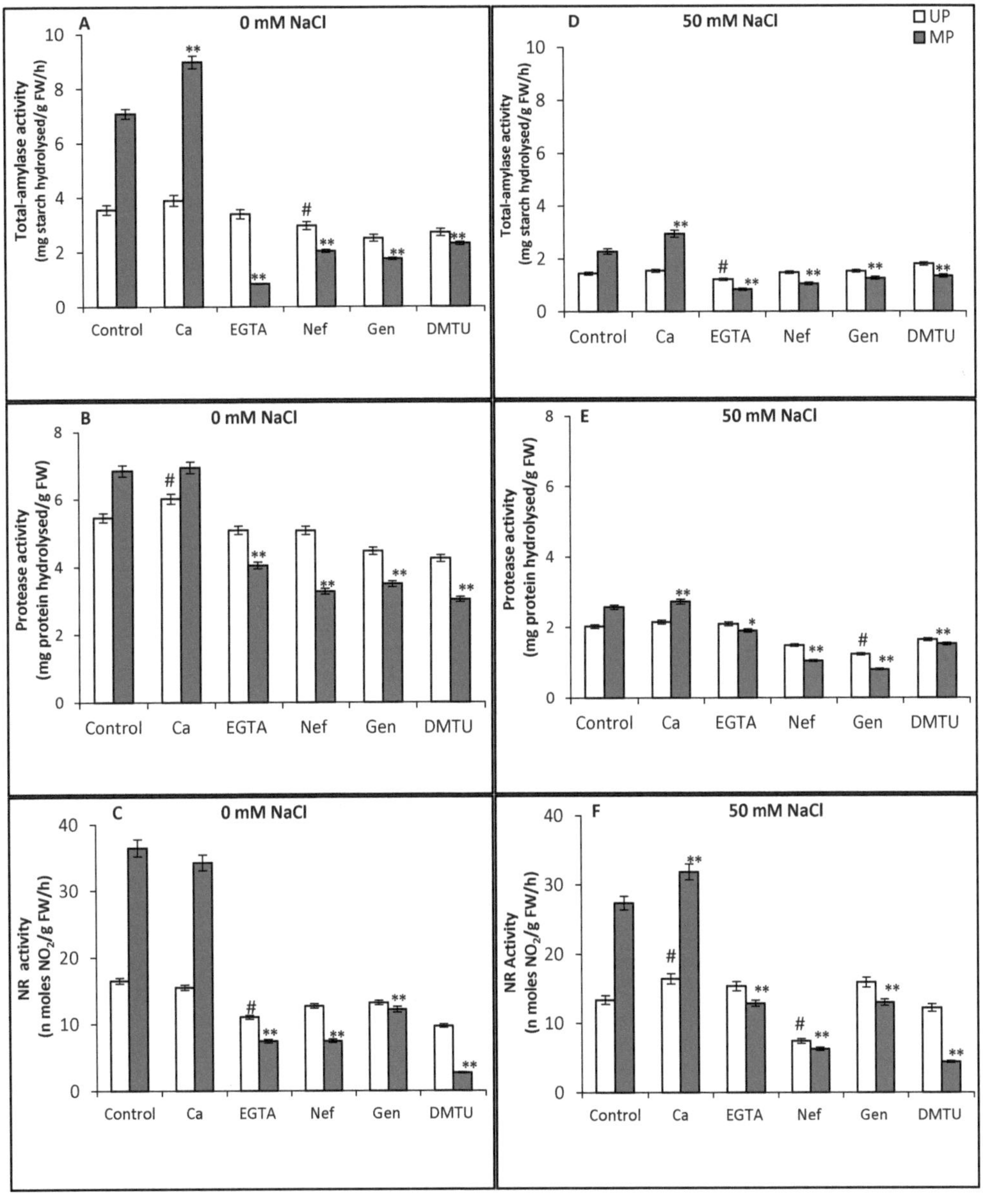

Figure 3. Effects of different modulators of calcium and cGMP on SMF pre-treatment (200 mT for 1 h) induced α-amylase (**A**,**D**), protease (**B**,**E**) and nitrate reductase (**C**,**F**) activity in soybean seedlings, respectively, in non-saline and saline conditions. These data are expressed as mean + SE. # $p < 0.05$ indicates the significant difference amongst the UP seeds grown in control with the UP seeds grown in different modulators under non-saline or saline conditions and ** $p < 0.01$, * $p < 0.05$ indicates the significant difference amongst the MP seeds grown in control with the MP grown in different modulators under non-saline or saline conditions. UP, un-primed seeds and MP, magneto-primed seeds. Ca, calcium chloride ($CaCl_2$), 1 mM (Ca^{2+} analogue); EGTA, ethylene glycol-β-amino ethyl ether tetra acetic acid, 0.1 mM; Nef, nifedipine, 1 mM (Ca^{2+} channel blocker); Gen, genistein, 0.05 mM (cGMP blocker), and dimethyl thiourea, 0.1 mM (DMTU, H_2O_2 inhibitor).

After SMF pre-treatment, the NO content in soybean seedlings was significantly enhanced in the absence (60%) and presence (82%) of salt stress (Figure 2C,F). This increase in NO content was maximally reduced by DMTU (80%), Nef (75%), EGTA (70%), and then Gen (67%) in non-saline conditions (Figure 2C). In presence of salt stress (50 mM NaCl) a reduction in NO content by 76, 50, 53, and 33% after EGTA, Nef, Gen and DMTU treatments, respectively, was observed (Figure 2F). On the other hand, in both saline and non-saline conditions the $CaCl_2$ significantly increased the H_2O_2 and NO contents in seedlings of both un-primed and SMF-primed seeds (Figure 2A–F).

$CaCl_2$ accelerated the total amylase activity of SMF-primed seeds by 27% and 29% in the absence and presence of salt stress conditions, respectively (Figure 3D). Maximum reductions of 63% (EGTA), 54% (Nef), 44% (Gen), and 41% (DMTU) were obtained in the total amylase activity of SMF-primed seeds in saline condition (Figure 3D). While un-primed seeds did not show any significant change by the modulators of Ca, cGMP, and ROS in non-saline as well as saline conditions (Figure 3A,D). The protease activity was enhanced by 25% and 27% in soybean seedlings from SMF-primed seeds as compared to un-primed seeds in the absence and presence of salinity, respectively (Figure 3B,E). The protease activity of SMF-primed seeds was declined by all the inhibitors of Ca^{2+}, cGMP and ROS (Figure 3B,E). The reduction in protease activity with Nef was 52% and 60% in SMF-primed seeds under non-saline (Figure 3B) and saline conditions, respectively (Figure 3E).

The NR activity in comparison to the un-primed seeds was remarkably enhanced in the soybean seedling during germination in SMF-primed seeds; an increase of 120 and 105% under the absence and presence of salt stress, respectively, was observed (Figure 3C,F). In the presence of salt stress, Ca^{2+}endorsed the activity of NR by 23% in un-primed seeds and 16% in SMF-primed seeds (Figure 3F). While blockers of Ca^{2+} and cGMP/H_2O_2 inhibitor DMTU inhibited the SMF-induced NR activity in non-saline as well as saline conditions (Figure 3C,F).

Previous studies have demonstrated that when plants are exposed to salinity stress, their growth is negatively impacted, but an increase in the concentration of Ca^{2+} in the cytosol can help alleviate these negative effects [53–55]. Likewise in our study by adding calcium to the experimental conditions, it showed positive effect on the studied parameters, potentially mitigating the negative impact of salinity in soybean. This is due to the fact that calcium functions as a second messenger in various biological systems, allowing plants to respond to high salt environments by activating a signal transduction pathway involving Ca^{2+} [56–58]. Seed priming can also alter the concentration of Ca^{2+}, which triggers a cascade of reactions involving calcium sensors such as calcium-dependent protein kinases (CDPKs) that regulate protein phosphorylation and gene expression during seed germination [59].

In this present study, various inhibitors of Ca^{2+}, cGMP, and ROS such as EGTA, Nef, Gen, and DMTU were tested on SMF-induced seedling growth under non-saline and saline conditions. The results showed that all the inhibitors significantly reduced SMF-induced seedling growth and vigour compared to un-primed seeds in both the conditions (Figure 1A–F). These findings suggest that SMF-priming may require the involvement of Ca^{2+} and cGMP in seed germination and seedling growth under salt stress. Previous studies have demonstrated that the addition of Ca^{2+} externally can mitigate the harmful effects of sodium on plant growth under hydroponic conditions, and the application of $Ca(NO_3)_2$ can enhance root nutrient and water uptake [60]. Additionally, research on *Brassica juncea* has shown that Ca^{2+} availability plays a crucial role in ABA-induced inhibition during seed germination, but does not affect the seed germination [43]. It was found that EGTA and nifedipine inhibited seed germination when tried individually and additionally raised the ABA effect, in a synergistic way, when tested simultaneously and concluded that Ca^{2+} is not essential for ABA to cause seed germination [43].

Previous research has shown that treating soybean seeds with SMF can improve the seed germination, which was attributed to an increase in the production of ROS [61]. During seed germination, the generation of free radicals is associated with a shift from a quiescent state to an active metabolic state, promoting faster germination [62,63]. High levels of ROS can act as signalling molecules, promoting rapid axis growth and mobilisation of reserve materials [64]. Additionally, ROS production during seed germination helps with processes such as cell wall elongation, endosperm weakening, redox regulation, hormone and Ca^{2+} signalling, protection against pathogens, and gene expression [65]. Our results showed that magnetoprimed seeds had a significant increase in ROS production, including $O_2^{\bullet-}$, H_2O_2, and NO content under both saline and non-saline conditions. Furthermore, the addition of Ca^{2+} further enhanced SMF-induced H_2O_2 and NO production, while the use of inhibitors such as EGTA, Nef, and DMTU significantly reduced ROS and NO production in both saline and non-saline conditions (Figure 2).

In addition, we found that magnetoprimed seeds exhibited higher levels of protease, amylase, and nitrate reductase activity compared to un-primed seeds under both saline and non-saline conditions (Figure 3A–F). Furthermore, the addition of Ca^{2+} further enhanced SMF-induced enzyme activities, while the use of inhibitors such as EGTA, Nef, and DMTU reduced the enzyme activities in both saline and non-saline conditions (Figure 3A–F). During seed germination, α and β-amylase enzymes results in the breakdown of stored carbohydrate in to the monosaccharides which are utilised by the growing seedlings. Previously, it has been reported that in soybean, chickpea, and maize under salinity stress there is increased rate of seed germination in magnetoprimed seeds because of higher activity of amylase, protease, and nitrate reductase [19,20,42,66]. Results clearly indicated that all the inhibitors of Ca/cGMP/ROS declined the NR activity under saline and non-saline conditions, and it was maximally repressed by Nif and DMTU in un-primed and SMF-primed seeds under both conditions (Figure 3C,F).

Our results suggest that SMF may activate Ca^{2+} receptors and target proteins to enhance [Ca^{2+}] cytosolic level and Ca^{2+} during seed germination may persuade the generation of H_2O_2 in salinity stress. Li et al. [67] reported that in *Arabidopsis thaliana* roots salt stress induces H_2O_2 accumulation in Ca^{2+}-dependent salt resistance pathway. It is also revealed that Ca^{2+} signalling induced NO accumulation through inducing H_2O_2 generation during stomatal closure of guard cells in *Arabidopsis* [68]. Moreover, a correlationship among H_2O_2, calcium-sensing receptor (CAS) and NO was found in Ca^{2+}-dependent guard cell signalling [69]. In the present study, we also found that the SMF-induced NO generation was suppressed by Ca^{2+} channel blockers (EGTA/Nef) and cGMP blockers (Genestein) indicating that Ca^{2+}/cGMP may mediate the effect of SMF on NO production. It was also shown that, Ca^{2+} released through various type of Ca^{2+} channels was activated by NO and H_2O_2 [70]. NO is a signalling molecule, with transduction through a cGMP-independent or cGMP-dependent pathway. In the cGMP-dependent pathway, NO signalling involves Ca^{2+},cGMP, cADPR, and protein kinases [71].

It has been demonstrated that the Ca^{2+} is the downstream targets of NO and it may act through cGMP and cADPR to control intra-cellular Ca^{2+} channels to increase free cytosolic calcium [72,73]. It was shown that NO was able to activate both intra-cellular and plasma membrane Ca^{2+} channels via cascades of reaction involving plasma membrane depolarization, CADPR, and protein kinase [72–74].The role of cGMP is strongly associated with the NO signal cascade in various physiological processes, such as seed germination and gibberellic acid(GA) induced α-amylase production [44,74,75]. Similarly, previously we found that magnetopriming-induced salt tolerance by NO signalling further activated GA synthesis and reduced the ABA content in soybean seeds during seed germination [22]. Thus, SMF exposure appears to stimulate a cascade of signalling events that culminate in the activation of key enzymes involved in seed germination and seedling growth. These findings suggest that SMF treatment may be a promising strategy for improving plant growth and yield in salt-affected soils. However, further research is needed to confirm these findings and optimize the SMF treatment conditions for maximum benefit.

4. Conclusions

In conclusion, it appears that exposure to SMF can activate calcium receptors and target proteins, leading to enhanced calcium and cGMP signalling. This signalling pathway may trigger the production of H_2O_2, O_2^-, and NO, which in turn may activate total amylase, protease, and nitrate reductase activities. Under salt stress conditions, the activation of these enzymes by SMF-priming may play an important role in accelerating seed germination and promoting the seedling growth in soybean plants. The addition of Ca^{2+} further enhanced SMF-induced growth and enzyme activities, while the use of inhibitors such as EGTA, Nef, Gen, and DMTU reduced all the parameters in both saline and non-saline conditions. Indeed, the increased Ca^{2+} resulting from SMF exposure may trigger the Ca^{2+}-mediated activation of the SOS signalling pathway, which plays a crucial role in maintaining ion homeostasis in plants under salt stress. These indicate that the SMF-induced tolerance against salt stress in soybean seedlings may be mediated, at least in part, by a complex interplay among Ca^{2+}, cGMP, and ROS signalling pathways. Further studies are needed to elucidate the precise mechanisms underlying this interplay and to identify specific molecular targets that can be manipulated to enhance plant tolerance to salt stress. This study could ultimately help to develop strategies for enhancing plant tolerance to abiotic stress, which is essential for sustainable agriculture in the face of climate change.

Author Contributions: S.K. and M.J. designed the experiments; S.S. and S.K. performed the experiments; S.K. and S.S. analysed the data; S.K. and M.J. wrote the manuscript; S.K., M.J., K.J., R.G. and S.S. finalised the manuscript in its final form. All authors have read and agreed to the published version of the manuscript.

Funding: This research work received financial support from the Department of Science Technology (SR/WOS-A/LS-17/2017-G) to S. Kataria.

Institutional Review Board Statement: Not applicable.

Informed Consent Statement: Not applicable.

Data Availability Statement: Data can be obtained on request from the authors.

Conflicts of Interest: The authors declare no conflict of interest.

References

1. Kataria, S.; Verma, S.K. Salinity stress responses and adaptive mechanisms in major glycophytic crops: The story so far. In *Salinity Responses and Tolerance in Plants: Targeting Sensory, Transport and Signaling Mechanisms*; Springer: Berlin/Heidelberg, Germany, 2018; Volume 1, pp. 1–39.
2. Kamran, M.; Parveen, A.; Ahmar, S.; Malik, Z.; Hussain, S.; Chattha, M.S.; Saleem, M.H.; Adil, M.; Heidari, P.; Chen, J.T. An overview of hazardous impacts of soil salinity in crops, tolerance mechanisms, and amelioration through selenium supplementation. *Int. J. Mol. Sci.* **2020**, *21*, 148. [CrossRef] [PubMed]
3. Tomar, R.S.; Kataria, S.; Jajoo, A. Behind the scene: Critical role of reactive oxygen species and reactive nitrogen species in salt stress tolerance. *J. Agron. Crop. Sci.* **2021**, *207*, 577–588. [CrossRef]
4. Ibrahim, E.A. Seed priming to alleviate salinity stress in germinating seeds. *J. Plant Physiol.* **2016**, *192*, 38–46. [CrossRef] [PubMed]
5. Venâncio, C.; Pereira, R.; Lopes, I. The influence of salinization on seed germination and plant growth under mono and polyculture. *Environ. Pollut.* **2020**, *260*, 113993. [CrossRef]
6. Hajihashemi, S.; Skalicky, M.; Brestic, M.; Pavla, V. Cross-talk between nitric oxide, hydrogen peroxide and calcium in salt-stressed *Chenopodium quinoa* Willdat seed germination stage. *Plant Physiol. Biochem.* **2020**, *154*, 657–664. [CrossRef]
7. Kubala, S.; Wojtyla, Ł.; Quinet, M.; Lechowska, K.; Lutts, S.; Garnczarska, M. Enhanced expression of the proline synthesis gene P5CSA in relation to seed osmopriming improvement of *Brassica napus* germination under salinity stress. *J. Plant Physiol.* **2015**, *183*, 1–12. [CrossRef] [PubMed]
8. Kataria, S.; Jain, M. Magnetopriming alleviates adverse effects of abiotic stresses in plants. In *Plant Tolerance to Environmental Stress*, 1st ed.; Hasanuzzaman, M., Fujita, M., Oku, H., Islam, T.M., Eds.; CRC Press: Boca Raton, FL, USA, 2019; pp. 427–442.
9. Waqas, M.; Korres, N.E.; Khan, M.D.; Nizami, A.S.; Deeba, F.; Ali, I.; Hussain, H. Advances in the concept and methods of seed priming. In *Priming and Pretreatment of Seeds and Seedlings: Implication in Plant Stress Tolerance and Enhancing Productivity in Crop Plants*; Hasanuzzaman, M., Fotopoulos, V., Eds.; Springer: Berlin/Heidelberg, Germany, 2019; pp. 11–41.
10. Prajapati, R.; Kataria, S.; Jain, M. Seed priming for alleviation of heavy metal toxicity in plants: An overview. *Plant Sci. Today* **2020**, *7*, 308–313. [CrossRef]

11. Sarraf, M.; Kataria, S.; Taimourya, H.; Santos, L.O.; Menegatti, R.D.; Jain, M.; Ihtisham, M.; Liu, S. Magnetic Field (MF) Applications in Plants: An Overview. *Plants* **2020**, *9*, 1139. [CrossRef]
12. Menegatti, R.D.; De Oliveira, L.O.; Da Costa, Á.V.L.; Braga, E.J.B.; Bianchi, V.J. Magnetic field and gibberellic acid as pre-germination treatments of passion fruit seeds. *Rev. Ciência Agrícola* **2019**, *17*, 15–22. [CrossRef]
13. Vashisth, A.; Nagarajan, S. Exposure of seeds to static magnetic field enhances germination and early growth characteristics in chickpea (*Cicer arietinum* L.). *BioElectroMagnetics* **2008**, *29*, 571–578. [CrossRef]
14. Bhardwaj, J.; Anand, A.; Nagarajan, S. Biochemical and biophysical changes associated with magnetopriming in germinating cucumber seeds. *Plant Physiol. Biochem.* **2012**, *57*, 67–73. [CrossRef] [PubMed]
15. Reina, F.G.; Pascual, L.A. Influence of a stationary magnetic field on water relations in lettuce seeds. Part I: Theoretical considerations. *BioElectroMagnetics* **2001**, *22*, 589–595. [CrossRef] [PubMed]
16. Vashisth, A.; Joshi, D.K. Growth characteristics of maize seeds exposed to magnetic field. *Bioelectromagnetics* **2017**, *38*, 151–157. [CrossRef]
17. Poinapen, D.; Brown, D.C.; Beeharry, G.K. Seed orientation and magnetic field strength have more influence on tomato seed performance than relative humidity and duration of exposure to non-uniform static magnetic fields. *J. Plant Physiol.* **2013**, *170*, 1251–1258. [CrossRef]
18. Konefał-Janocha, M.; Banaś-Ząbczyk, A.; Bester, M.; Bocak, D.; Budzik, S.; Górny, S.; Cholewa, M. The effect of stationary and variable electromagnetic fields on the germination and early growth of radish (*Raphanussativus*). *Pol. J. Environ. Stud.* **2018**, *28*, 709–715. [CrossRef] [PubMed]
19. Kataria, S.; Baghel, L.; Guruprasad, K. Pre-treatment of seeds with static magnetic field improves germination and early growth characteristics under salt stress in maize and soybean. *Biocatal. Agric. Biotechnol.* **2017**, *10*, 83–90. [CrossRef]
20. Thomas, S.; Anand, A.; Chinnusamy, V.; Dahuja, A.; Basu, S. Magnetopriming circumvents the effect of salinity stress on germination in chickpea seeds. *Acta Physiol. Plant.* **2013**, *35*, 3401–3411. [CrossRef]
21. Kataria, S.; Jain, M.; Tripathi, D.K.; Singh, V.P. Involvement of nitrate reductase-dependent nitric oxide production in magnetopriming-induced salt tolerance in soybean. *Physiol. Plant.* **2020**, *168*, 422–436. [CrossRef]
22. Kataria, S.; Anand, A.; Raipuria, R.K.; Kumar, S.; Jain, M.; Watts, A.; Brestic, M. Magnetopriming actuates nitric oxide synthesis to regulate phytohormones for improving germination of soybean seeds under salt stress. *Cells* **2022**, *11*, 2174. [CrossRef]
23. Jumrani, K.; Bhatia, V.S.; Kataria, S.; Alamri, S.; Siddiqui, M.H.; Rastogi, A. Inoculation with arbuscular mycorrhizal fungi alleviates the adverse effects of high temperature in soybean. *Plants* **2022**, *11*, 2210. [CrossRef]
24. Kataria, S.; Baghel, L.; Jain, M.; Guruprasad, K.N. Magnetopriming regulates antioxidant defense system in soybean against salt stress. *Biocatal. Agric. Biotechnol.* **2019**, *18*, 101090. [CrossRef]
25. De Souza, A.; Garcí, D.; Sueiro, L.; Gilart, F.; Porras, E.; Licea, L. Pre-sowing magnetic treatments of tomato seeds increase the growth and yield of plants. *Bioelectromagnetics* **2006**, *4*, 247–257. [CrossRef] [PubMed]
26. Chen, Y.P.; Li, R.; He, J.M. Magnetic field can alleviate toxicological effect induced by cadmium in mungbean seedlings. *Ecotoxicology* **2011**, *20*, 760–769. [CrossRef] [PubMed]
27. Baghel, L.; Kataria, S.; Guruprasad, K.N. Static magnetic field treatment of seeds improves carbon and nitrogen metabolism under salinity stress in soybean. *Bioelectromagnetics* **2016**, *37*, 455–470. [CrossRef] [PubMed]
28. Baghel, L.; Kataria, S.; Guruprasad, K.N. Effect of static magnetic field pre-treatment on growth, photosynthetic performance and yield of soybean under water stress. *Photosynthetica* **2018**, *56*, 718–730. [CrossRef]
29. Hozayn, M.; EL-Mahdy, A.A.; Zalama, M.T. Magneto-priming for improving germination, seedling attributes and field performance of barley (*Hordeum vulgare* L.) under salinity stress. *Middle East J. Agric Res.* **2018**, *7*, 1006–1022.
30. Kataria, S.; Jain, M.; Rastogi, A.; Brestic, M. Static magnetic field treatment enhanced photosynthetic performance in soybean under supplemental ultraviolet-B radiation. *Photosynth. Res.* **2021**, *150*, 263–278. [CrossRef]
31. Aliniaeifard, S.; Shomali, A.; Seifikalhor, M.; Lastochkina, O. Calcium signalling in plants under drought. In *Salt and Drought Stress Tolerance in Plants*; Springer: Berlin/Heidelberg, Germany, 2020; pp. 259–298.
32. Seifikalhor, M.; Aliniaeifard, S.; Shomali, A.; Azad, N.; Hassani, B.; Lastochkina, O.; Li, T. Calcium signaling and salt tolerance are diversely entwined in plants. *Plant Signal. Behav.* **2019**, *14*, 1665455. [CrossRef]
33. Kudla, J.; Batistič, O.; Hashimoto, K. Calcium signals: The lead currency of plant information processing. *Plant Cell* **2010**, *22*, 541–563. [CrossRef]
34. Bakeer, S.M. Effect of ammonium nitrate fertilizer and calcium chloride foliar spray on fruit cracking and sunburn of *Manflouty pomegranate* trees. *Sci. Hortic.* **2016**, *209*, 300–308. [CrossRef]
35. Choi, W.G.; Hilleary, R.; Swanson, S.J.; Kim, S.H.; Gilroy, S. Rapid, long-distance electrical and calcium signaling in plants. *Annu. Rev. Plant Biol.* **2016**, *67*, 287–307. [CrossRef] [PubMed]
36. Xu, D.H.; Wang, W.; Gao, T.; Fang, X.; Gao, X.; Li, J.; Bu, H.; Mu, J. Calcium alleviates decreases in photosynthesis under salt stress by enhancing antioxidant metabolism and adjusting solute accumulation in *Calligonum mongolicum*. *Conserv. Physiol.* **2017**, *5*, cox060. [CrossRef]
37. Xu, T.; Niu, J.; Jiang, Z. Sensing Mechanisms: Calcium Signaling Mediated Abiotic Stress in Plants. *Front. Plant Sci.* **2022**, *13*, 925863. [CrossRef] [PubMed]
38. Niu, L.; Liao, W. Hydrogen peroxide signaling in plant development and abiotic responses: Crosstalk with nitric oxide and calcium. *Front. Plant Sci.* **2016**, *4*, 230. [CrossRef] [PubMed]

39. Hasanuzzaman, M.; Oku, H.; Nahar, K.; Bhuyan, M.B.; Al Mahmud, J.; Baluska, F.; Fujita, M. Nitric oxide-induced salt stress tolerance in plants: ROS metabolism, signaling and molecular interactions. *Plant Biotech. Rep.* **2018**, *12*, 77–92. [CrossRef]
40. Gohari, G.; Alavi, Z.; Esfandiari, E.; Panahirad, S.; Hajihoseinlou, S.; Fotopoulos, V. Interaction between hydrogen peroxide and sodium nitroprusside following chemical priming of *Ocimum basilicum* L. against salt stress. *Physiol. Plant.* **2020**, *168*, 361–373.
41. Liu, L.; Huang, L.; Sun, C.; Wang, L.; Jin, C.; Lin, X. Cross-talk between hydrogen peroxide and nitric oxide during plant development and responses to stress. *J. Agric. Food Chem.* **2021**, 33, 9485–9497. [CrossRef]
42. Kataria, S.; Baghel, L.; Guruprasad, K.N. Alleviation of adverse effects of ambient UV stress on growth and some potential physiological attributes in soybean (*Glycine max*) by seed pre-treatment with static magnetic field. *J. Plant Growth Regul.* **2017**, *36*, 550–565. [CrossRef]
43. Abdul-Baki, A.A.; Anderson, J.D. Vigour determination in soybean by multiple criteria. *Crop. Sci.* **1973**, *10*, 31–34.
44. Sharma, S.S.; Sharma, S.; Rai, V.K. The effect of EGTA, calcium channel blockers (Lanthanum Chloride and Nifedipine) and their interaction with abscisic acid on seed germination of *Brassica juncea* cv. RLM-198. *Ann. Bot.* **1992**, *70*, 295–299. [CrossRef]
45. Teng, Y.; Xu, W.; Ma, M. cGMP is required for seed germination in *Arabidopsis thaliana*. *J. Plant Physiol.* **2010**, *167*, 885–889. [CrossRef] [PubMed]
46. Deng, B.; Kejun, Y.; Zhang, Y.; Zuotong, L. Can heavy metal pollution defend seed germination against heat stress? Effect of heavy metals (Cu^{2+}, Cd^{2+} and Hg^{2+}) on maize seed germination under high temperature. *Environ. Poll.* **2016**, *216*, 46–52. [CrossRef] [PubMed]
47. Sawhney, S.; Toky, K.L.; Nanda, K.K. Changes in amylase activity during extension growth and floral induction in *Impatiens balsamina* a qualitative short day plant. *Indian J. Plant Physiol.* **1970**, *13*, 198–205.
48. Kunitz, M. Crystalline soybean trypsin inhibitor: II. General properties. *J. Gen. Physiol.* **1947**, *30*, 291–300. [CrossRef] [PubMed]
49. Jaworski, E.G. Nitrate reductase assay in intact plant tissues. *Biochem. Biophys. Res. Commun.* **1971**, *43*, 1274–1279. [CrossRef]
50. Zhu, J.K. Regulation of ion homeostasis under salt stress. *Curr. Opin. Plant Biol.* **2003**, *6*, 441–445. [CrossRef]
51. Chaitanya, K.S.K.; Naithani, S.C. Role of superoxide, lipid peroxidation and superoxide dismutase in membrane perturbations during loss of viability in seeds of *Shorea robusta* (Gaertn. F.). *New Phytol.* **1994**, *126*, 623–627. [CrossRef]
52. Mukherjee, S.P.; Choudhari, M.A. Implications of water stressed induced changes in the levels of endogenous ascorbic acid and H_2O_2 in Vigna seedlings. *Physiol. Plant.* **1983**, *58*, 166–170. [CrossRef]
53. Tuna, A.L.; Kaya, C.; Ashraf, M.; Altunlu, H.; Yokas, I.; Yagmur, B. The effects of calcium sulphate on growth, membrane stability and nutrient uptake of tomato plants grown under salt stress. *Environ. Exp. Bot.* **2007**, *59*, 173–178. [CrossRef]
54. Marschner, P.; Rengel, Z. Nutrient availability in soils. In *Marschner's Mineral Nutrition of Higher Plants*; Academic Press: Cambridge, MA, USA, 2012; pp. 315–330.
55. Murillo-Amador, B.; Troyo-Diéguez, E.; García-Hernández, J.L.; López-Aguilar, R.; Ávila-Serrano, N.Y.; Zamora-Salgado, S.; Rueda-Puente, E.O.; Kaya, C. Effect of NaCl salinity in the genotypic variation of cowpea (*Vigna unguiculata*) during early vegetative growth. *Sci. Hort.* **2006**, *108*, 423–431. [CrossRef]
56. Hasegawa, P.M.; Bressan, R.A.; Zhu, J.K.; Bohnert, H.J. Plant cellular and molecular responses to high salinity. *Annu. Rev. Plant Physiol. Plant Mol. Biol.* **2000**, *51*, 463–499. [CrossRef]
57. Scrase-Field, S.A.; Knight, M.R. Calcium: Just a chemical switch? *Curr. Opin. Plant Biol.* **2003**, *6*, 500–506. [CrossRef] [PubMed]
58. Parre, E.; Ghars, M.A.; Leprince, A.S.; Thiery, L.; Lefebvre, D.; Bordenave, M.; Richard, L.; Mazars, C.; Abdelly, C.; Savouré, A. Calcium signaling via phospholipase C is essential for proline accumulation upon ionic but not nonionic hyperosmotic stresses in Arabidopsis. *Plant Physiol.* **2007**, *144*, 503–512. [CrossRef] [PubMed]
59. Anil, V.S.; Sankara, K.R. Calcium-mediated signaling during sandalwood somatic embryogenesis. Role for exogenous calcium as second messenger. *Plant Physiol.* **2000**, *123*, 1301–1311. [CrossRef]
60. Patel, N.T.; Vaghela, P.M.; Patel, A.D.; Pandey, A.N. Implications of calcium nutrition on the response of *Caesalpinia crista* (*Fabaceae*) to soil salinity. *Acta Ecol. Sin.* **2011**, *31*, 24–30. [CrossRef]
61. Shine, M.B.; Guruprasad, K.N.; Anand, A. Effect of stationary magnetic field strengths of 150 and 200 mT on reactive oxygen species production in soybean. *BioElectroMagnetics* **2012**, *33*, 428–437. [CrossRef]
62. Gidrol, X.; Lin, W.S.; Dégousée, N.; Yip, S.F.; Kush, A. Accumulation of reactive oxygen species and oxidation of cytokinin in germinating soybean seeds. *Eur. J. Biochem.* **1994**, *224*, 21–28. [CrossRef]
63. Boucelha, L.; Djebbar, R.; Abrous-Belbachir, O. *Vigna unguiculata* seed priming is related to redox status of plumule, radicle and cotyledons. *Funct. Plant Biol.* **2019**, *46*, 584–594. [CrossRef]
64. Sachdev, S.; Ansari, S.A.; Ansari, M.I.; Fujita, M.; Hasanuzzaman, M. Abiotic stress and reactive oxygen species: Generation, signaling, and defense mechanisms. *Antioxidants* **2021**, *10*, 277. [CrossRef]
65. El-Maarouf-Bouteau, H.; Bailly, C. Oxidative signaling in seed germination and dormancy. *Plant Signal. Behav.* **2008**, *3*, 175–182. [CrossRef]
66. Baghel, L.; Kataria, S.; Jain, M. Mitigation of adverse effects of salt stress on germination, growth, photosynthetic efficiency and yield in maize (*Zea mays* L.) through magnetopriming. *Acta Agrobot.* **2019**, *72*, 1757. [CrossRef]
67. Li, J.; Wang, X.; Zhang, Y.; Jia, H.; Bi, Y. cGMP regulates hydrogen peroxide accumulation in calcium-dependent salt resistance pathway in *Arabidopsis thaliana* roots. *Planta* **2011**, *234*, 709–722. [CrossRef]

68. Li, J.-H.; Liu, Y.-Q.; Lu, P.; Lin, H.-F.; Bai, Y.; Wang, X.-C.; Chen, Y.-L. A signaling pathway linking nitric oxide production to heterotrimeric g protein and hydrogen peroxide regulates extracellular calmodulin induction of stomatal closure in *Arabidopsis*. *Plant Physiol.* **2009**, *150*, 114–124. [CrossRef]
69. Wang, W.H.; Yi, X.Q.; Han, A.D.; Liu, T.W.; Chen, J.; Wu, F.H.; Dong, X.J.; He, J.X.; Pei, Z.M.; Zheng, H.L. Calcium-sensing receptor regulates stomatal closure through hydrogen peroxide and nitric oxide in response to extracellular calcium in *Arabidopsis*. *J. Exp. Bot.* **2012**, *63*, 177–190. [CrossRef] [PubMed]
70. González, A.; Cabrera, M.A.; Henríquez, M.J.; Contreras, R.A.; Morales, B.; Moenne, A. Cross talk among calcium, hydrogen peroxide, and nitric oxide and activation of gene expression involving calmodulins and calcium-dependent protein kinases in *Ulva compressa* exposed to copper excess. *Plant Physiol.* **2012**, *158*, 1451–1462. [CrossRef] [PubMed]
71. Magdalena, A.; Jolanta, F.W. Nitric oxide as a bioactive signaling molecule in plant stress responses. *Plant Sci.* **2007**, *172*, 876–887.
72. Wendehenne, D.; Durner, J.; Klessig, D.F. Nitric oxide: A new player in plant signaling and defence responses. *Curr. Opin. Plant Biol.* **2004**, *7*, 449–455. [CrossRef]
73. Lamotte, O.; Gould, K.; Lecourieux, D.; Sequeira-Legrand, A.; Lebrun-Garcia, A.; Durner, J.; Pugin, A.; Wendehenne, D. Analysis of nitric oxide signalling functions in tobacco cells challenged by the elicitor cryptogein. *Plant Physiol.* **2004**, *135*, 516–529. [CrossRef]
74. Penson, S.P.; Schuurink, R.C.; Fath, A.; Gubler, F.; Jacobsen, J.V.; Jones, R.L. cGMP is required for gibberellic acid-induced gene expression in barley aleurone. *Plant Cell* **1996**, *8*, 2325–2333. [CrossRef]
75. Wu, M.; Wang, F.; Zhang, C.; Xie, Y.; Han, B.; Huang, J.; Shen, W. Heme oxygenase-1 is involved in nitric oxide and cGMP-induced α-Amy2/54 gene expression in GA-treated wheat aleurone layers. *Plant Mol. Biol.* **2013**, *81*, 27–40. [CrossRef]

seeds

MDPI

Article

Can Osmopriming Induce Cross-Tolerance for Abiotic Stresses in *Solanum paniculatum* L. Seeds? A Transcriptome Analysis Point of View

Pedro Bento da Silva [1], Tatiana Arantes Afonso Vaz [2,3,*], Marcio Luis Acencio [4,†], Luiz Augusto Bovolenta [4], Henk W. M. Hilhorst [5] and Edvaldo A. Amaral da Silva [1]

1 Departamento de Produção Vegetal, Universidade Estadual Paulista "Júlio de Mesquita Filho", FCA-UNESP, Botucatu 18610-034, SP, Brazil; pb.bento@gmail.com (P.B.d.S.); amaral.silva@unesp.br (E.A.A.d.S.)
2 Central de Laboratórios, Universidade Federal do Triângulo Mineiro, UFTM, Uberaba 38025-180, MG, Brazil
3 Laboratório de Sementes Florestais, Universidade Federal de Lavras, UFLA, Lavras 37200-000, MG, Brazil
4 Departamento de Biofísica e Farmacologia, Universidade Estadual Paulista "Júlio de Mesquita Filho", IBB-UNESP, Botucatu 18610-034, SP, Brazil; mlacencio@gmail.com (M.L.A.); labovolenta@gmail.com (L.A.B.)
5 Laboratory of Plant Physiology, Wageningen University, 6708 PB Wageningen, The Netherlands; henk.hilhorst@wur.nl
* Correspondence: tatiana.arantes@gmail.com
† Current address: Luxembourg Centre for Systems Biomedicine, University of Luxembourg, 4362 Esch-sur-Alzette, Luxembourg.

Abstract: *Solanum paniculatum* L. belongs to the Solanaceae family and has the ability to grow and develop under unfavorable environmental conditions such as drought and salt stress, acid soils and soils poor in nutrients. The present work aimed to analyze *S. paniculatum* seed transcriptome associated with induced tolerance to drought stress by osmopriming. Seeds subjected to osmopriming (−1.0 MPa) displayed a higher germination and normal seedling percentage under drought stress when compared with unprimed seeds. RNA-seq transcriptome profiles of osmoprimed and unprimed seeds were determined and the potential proteins involved in the drought tolerance of *S. paniculatum* were identified. From the 34,640 assembled transcripts for both osmoprimed and unprimed seeds, only 235 were differentially expressed and, among these, 23 (10%) transcripts were predicted to code for proteins potentially involved in response to stress, response to abiotic stimulus and response to chemical. The possible mechanisms by which these stress-associated genes may confer tolerance to osmoprimed *Solanum paniculatum* seeds to germinate under water deficit was discussed and may help to find markers for the selection of new materials belonging to the Solanaceae family that are more tolerant to stress during and following germination.

Keywords: differential expression; RNA-Seq; priming; stress tolerance; jurubeba

Citation: da Silva, P.B.; Vaz, T.A.A.; Acencio, M.L.; Bovolenta, L.A.; Hilhorst, H.W.M.; da Silva, E.A.A. Can Osmopriming Induce Cross-Tolerance for Abiotic Stresses in *Solanum paniculatum* L. Seeds? A Transcriptome Analysis Point of View. *Seeds* **2023**, *2*, 382–393. https://doi.org/10.3390/seeds2040029

Academic Editors: José Antonio Hernández Cortés, Gregorio Barba-Espín and Pedro Diaz-Vivancos

Received: 21 July 2023
Revised: 8 September 2023
Accepted: 14 September 2023
Published: 28 September 2023

1. Introduction

Successful plant establishment relies on seed quality and its ability to overcome challenges during the germination process. Throughout their life cycle, plants are exposed to biotic and abiotic stresses [1] and experience the most sensitive stage as seedlings. *Solanum paniculatum* L. belongs to the Solanaceae family and has a remarkable capacity to grow and develop under unfavorable conditions such as drought, acidic soils and soils poor in nutrients [2–4].

In the genus *Solanum*, there are species with a high degree of representativeness due to their economic importance such as potato (*S. tuberosum* L.), tomato (*S. lycopersicum* L.), eggplant (*S. melongena* L.) and others. Also, this genus contains ornamental and weed species [5]. Wild relative species can show higher thresholds of stress tolerance than crops and knowing their physiology can be useful for the development of quality markers and for the production of plant materials that are tolerant to environmental stresses [6–8].

Kranner [9] classifies as eustress the conditions that result in positive changes to plant metabolism, and also to seed performance. The principle of cross-tolerance is based on the assumption that the exposure of seeds (or plants) to one type of stress induces a common tolerance to various abiotic and biotic stresses and that the activity of the genes and/or proteins involved in this response is preserved among the seed, seedling and plant [10–13]. Seed priming premises are based on the cross-tolerance concept and are well known for improving germination percentage, germination uniformity, seed quality and seedling establishment under adverse conditions [14–16].

During priming, seed hydration is controlled using osmotic solutions, time and/or temperature of imbibition (and other techniques) in order to stop the germination process at phase II, so radicle protrusion (phase III) is prevented [17]. Osmopriming consists of providing an aerated solution containing sugars or polyethylene glycol (PEG 6000 or 8000) to seeds [18,19] and incubating them in suboptimal temperature conditions. Therefore, the mild stress caused by priming may induce stress tolerance as well as the improvement of seed lot quality [16,19,20].

Seed priming and subsequent transcriptomic analysis using RNAseq technology can be used to understand the molecular mechanisms developed by *S. paniculatum* to grow under adverse conditions. In addition, it provides the identification of key genes involved in the stress response of economically important species of *Solanum* genus. Therefore, the main goal of this work was to analyze changes in the transcriptome of *S. paniculatum* seeds in response to osmopriming, aiming to identify potential proteins that confer stress tolerance.

2. Materials and Methods

Ripe fruits of *S. paniculatum* were collected from five plants in the State of Minas Gerais, Brazil (latitude 21°14′43″ S, longitude 44°59′59″ W and 918 m of altitude). Pulp was manually removed and passed through a sieve under running water to separate the seeds. After processing, the seeds were blotted dry with a paper towel and dried in a climate-controlled room (23 ± 2 °C, 60% RH) until they reached 8% moisture content (fresh weight basis). Thereafter, the seeds were stored in plastic bags in a cold room at 4 ± 2 °C until the beginning of the experiments.

To determine the seed water content, the seeds were submitted to artificial drying using four replicates of 0.5 g each placed in an oven at 103 ± 2 °C for 17 h, according to the International Seed Testing Association (ISTA) [21].

Seed imbibition was conducted using three replicates of 0.1 g of seeds each, placed in Petri dishes containing two layers of germination paper moistened with distilled water and incubated at 25 °C for a 12 h photoperiod in a germination chamber. The seeds were then removed from the Petri dishes, blotted dry with a paper towel and weighed on a precision scale (0.0001 g) every 30 min until two hours of imbibition and every two hours until 12 h. From there, the seeds were weighed every 24 h until radicle protrusion.

For the germination tests, four replicates of 25 seeds each were sterilized in a 1% sodium hypochlorite solution for 10 min then washed in running water for three minutes and blotted dry with a paper towel. The seeds were placed in 9 cm Petri dishes containing two layers of germination paper moistened with 5 mL of distilled water and incubated at 25 °C under constant light in a germination chamber. Germination was assessed daily for 17 days using the primary root length ≥1 mm as the criterion for germination.

For seed priming (hereafter named osmopriming), the seeds were immersed in 15 mL of polyethylene glycol (PEG 8000) solution of osmotic potentials of −0.4, −0.8, −1.0 and −1.2 MPa (prepared according [22,23]) and incubated at 15 °C for 15 days under constant light in a germination chamber. To avoid anoxia during incubation, small holes were made in the tube caps and tubes were placed in a shaker to favor aeration (Multifunctional mixer MR-II model-Biomixer). The PEG 8000 solutions were renewed after 24 h, 5 and 10 days of incubation.

Osmoprimed and control seeds (unprimed) were submitted to germination under stress conditions aiming to check the capacity of the treatments to induce tolerance to water deficit. After osmopriming, the seeds were washed in running water for one minute and placed in 9 cm Petri dishes containing two layers of germination paper moistened with 5 mL of distilled water (0.0 MPa) or PEG 8000 solution at osmotic potentials of −0.2, −0.4, −0.6, −0.8 and −1.0 MPa, then incubated at 25 °C under constant light. The tests were assessed daily for 30 days, the germination criteria was primary root ≥1 mm and normal seedlings were considered when presenting all primary structures (primary root, epicotyl, cotyledons and plumule). Aiming to maintain the osmotic potentials stable, the PEG solutions and substrates (germination paper) were changed every three days during the tests.

Data collected during germination in water at 25 °C were submitted to regression analysis with SigmaPlot v. 12.0. Germination data collected after osmopriming and under stress were submitted to Anova followed by means comparison using Tukey's test (p-value < 0.05) with SISVAR 5.4 (Tables S1 and S2).

Seeds submitted to −1.0 MPa osmopriming treatment showed better performance under water stress so they were chosen for the transcriptome analysis. Total RNA extraction was performed using 100 seeds that had been subjected to priming at −1.0 MPa and from 100 seeds subjected to 24 h of water imbibition (control). Total RNA was extracted using the NucleoSpin RNA Plant® kit (Macherey-Nagel, Dueren, Germany) according to the manufacturer's instructions. Total RNA from high-quality samples (RIN values > 7.0 evaluated by a 2100 Bioanalyzer, Agilent Technologies, Santa Clara, EUA) were used for library construction using the TruSeq RNA sample prep protocol v2 (Illumina, San Diego, EUA). The samples were sequenced using commercially available kits and HiScan platform (Illumina, San Diego, EUA) sequencing equipment, using a 50 bp single run module. All these steps were carried out strictly following the directions proposed by the manufacturer of the sequencing equipment. The data were analyzed by the CLC Genomics Workbench Version 6.0.2 (Bio CLC, Aarhus, Denmark) with default parameters for trimming, transcriptome assembly (de novo) and transcript quantification using the reads per kilobase per million of mapped reads (RPKM) normalization. The genes differentially expressed were generated by using Baggerley's test of samples [24]. Genes with p-value = 0 were selected for gene ontology analysis.

The functional annotation of transcripts was performed in three steps: (1) determination of the coding potential of transcripts, (2) selection of meaningful descriptions for the novel transcripts and (3) assignment of gene ontology (GO) terms (The Gene Ontology Consortium). The coding potential of all transcripts were determined using the support vector machine-based classifier Portrait, a software tool for non-coding RNA screening in transcriptome from poorly characterized species [25]. Only the transcripts predicted as coding by Portrait were retained for the next steps. For the selection of brief functional descriptions for the novel transcripts and the assignment of GO terms, we used the Blast2GO (Java WebStart version 2.8) [26] with default parameters. In brief, Blast2GO first uses BLASTX to find proteins in the NCBI NR database that are similar to the potential proteins encoded by the transcripts and then transfers both the brief functional descriptions and the GO terms from the most similar proteins to the novel transcripts. It is noteworthy to mention that (1) these functional descriptions are only exploratory in nature and that (2) more than one transcript can be annotated with the same functional description simply due to the fact that these transcripts may code, for example, for proteins belonging to the same protein family.

To confirm the results of RNA-seq data, differentially expressed transcripts (DETs) were used for primer design. In addition, the most stable transcripts for primed and unprimed seeds were selected to normalize the RT-qPCR data. The stable transcripts used were the Contig34 (cytochrome P450 87A3-like), Contig327 (Heat shock cognate 70 kDa protein 2-like) and Contig416 (Subtilisin-like protease-like). The software PerlPrimer (v1.1.21) was used for primer design with the following parameters: amplicon of 100 to 200 base pairs, annealing temperature of 60 °C ± 1 and base pairs varying from 20 to 24.

Gene expression was quantified by RT-qPCR using three biological replicates of 100 osmoprimed and 100 control (unprimed) seeds. The seeds were again placed in plastic tubes of 15 mL at −1.0 MPa of PEG solution at 15 °C for 15 days and total RNA was extracted as previously described. cDNA synthesis (reverse transcription) was performed by using the High Capacity cDNA Reverse Transcription kit (Applied Biosystems, Waltham, EUA) following the manufacturer's instructions. The real time PCR reactions were performed on a thermocycler (Eco Real-Time, Illumina, San Diego, EUA) with CtSybr Green qPCR Ready Mix (Sigma Life Science, Bath, UK) using the manufacturer's instructions and the primer efficiencies were assessed by LinReg PCR (v.11.0. Amsterdan, The Netherlands) software [27]. The data were analyzed by the EcoStudy program version 5.0 (Illumina, San Diego, EUA). Three biological replicates and three technical replicates for each sample were used. The gene expression data obtained by RT-qPCR were analyzed using the software Rest® 2009 (Qiagen, Munich, Germany), significant differences (p-value < 0.05) were considered as compared to the control. SigmaPlot® (Palo Alto, EUA) for Windows version 11 and 14.5 were used to prepare the artwork presented in this manuscript.

3. Results

Solanum paniculatum seeds were totally hydrated for between five and ten hours of imbibition (Figure 1A). Radicle protrusion started at the 7th day, took 13 days to reach 50% and seed germination reached 99% in 17 days (Figure 1B). As expected, osmoprimed seeds showed faster germination as compared with the control, with 50% germination occurring at 5 days and 99% of the seeds having germinated after 13 days (Figure 1B). Seeds submitted to water stress during germination showed a decrease in germinability and in the percentage of normal seedlings as the stress became higher, but, as expected, osmoprimed seeds had a better performance than seeds from the control group (Figure 1C). Seeds osmoprimed at −1.0 and −1.2 MPa could maintain higher germination and percentage of normal seedlings under moderate water stress (−0.4 MPa), standing out from the other treatments. So, −1.0 MPa osmopriming treatment was used to investigate changes in the transcriptome of *S. paniculatum* seeds.

For the cDNA libraries derived from osmoprimed and control seeds, 19,779,709 and 15,773,140 single-end reads, respectively, were generated. After the cleaning and removal of low-quality reads (phred scores < 20), 19,468,919 and 15,534,731 clean reads were identified, respectively, derived from osmoprimed and control seeds. Based on the high-quality reads, 34,640 contigs were assembled with sizes ranging from 249 to 5000 base pairs that were grouped into eight classes based on the number of base pairs (Figure 2). The largest class was the one with fragment lengths of 250–500 bp (39.1%).

From the total amount of transcripts generated (n = 34,640), 54.79% (n = 18,981) were shown to be coding and 45.20% (n = 15,659) were shown to be non-coding sequences. Of the 18,981 transcripts that encode some protein, only 25 (0.08%) show no similarity to any other sequence at the protein level. Of the remainder, 18,956 protein-coding transcripts showed similarity to protein sequences in the NCBI NR database, so it was possible to assign GO terms for 13,678 transcripts (73%) and it was not possible to assign any GO term for the remaining 5070 (27%). The reads are available in the NCBI Sequence Read Archive, under accession numbers STUDY: PRJNA384240 (SRP105294) SAMPLE: SRX2766227: RNA-seq of *Solanum paniculatum*: unprimed seeds and SAMPLE: SRX2766226: RNA-seq of *Solanum paniculatum*: primed seeds.

Among the 34,640 transcripts generated, 235 were differentially expressed (Table S3) between seeds subjected to osmopriming at −1.0 MPa and seeds subjected to 24 h of water imbibition (control), among which 232 were upregulated and three were downregulated. Of the 235 transcripts, 162 (69%) were potential protein-coding and 73 (31%) were potential non-coding transcripts. In order to confirm the accuracy of the RNA-Seq data, specific primers were designed for the transcripts of interest, i.e., highest expressed transcripts in osmoprimed seeds known to be involved in drought stress as well as for the stable transcripts used as a reference (Table 1). RT-qPCR results were consistent with the expression levels obtained by RNA-Seq (Figure 3). Interestingly, the most differentially expressed transcript was a potential non-coding transcript that was about 80-fold upregulated in osmoprimed seeds.

In addition to the 13 DETs known to be involved in drought stress (Table 1), we also sought to identify DETs coding for potential proteins that, according to GO annotations, could be involved in drought tolerance promoted by osmopriming. For this reason, protein-coding DETs were selected and functionally categorized in GO terms related to response to stress, chemical and abiotic stimulus (Table 2 and Table S2). Among the 235 DETs, 30 (13%)—including seven DETs previously identified as involved in drought stress (Table 1)—were associated with GO terms related to response to at least one of the above-mentioned specific stimuli. While 20 DETs were associated with GO terms related to response to stress, 19 and 15 DETs were associated, respectively, with response to chemical and abiotic stimulus (Table 2).

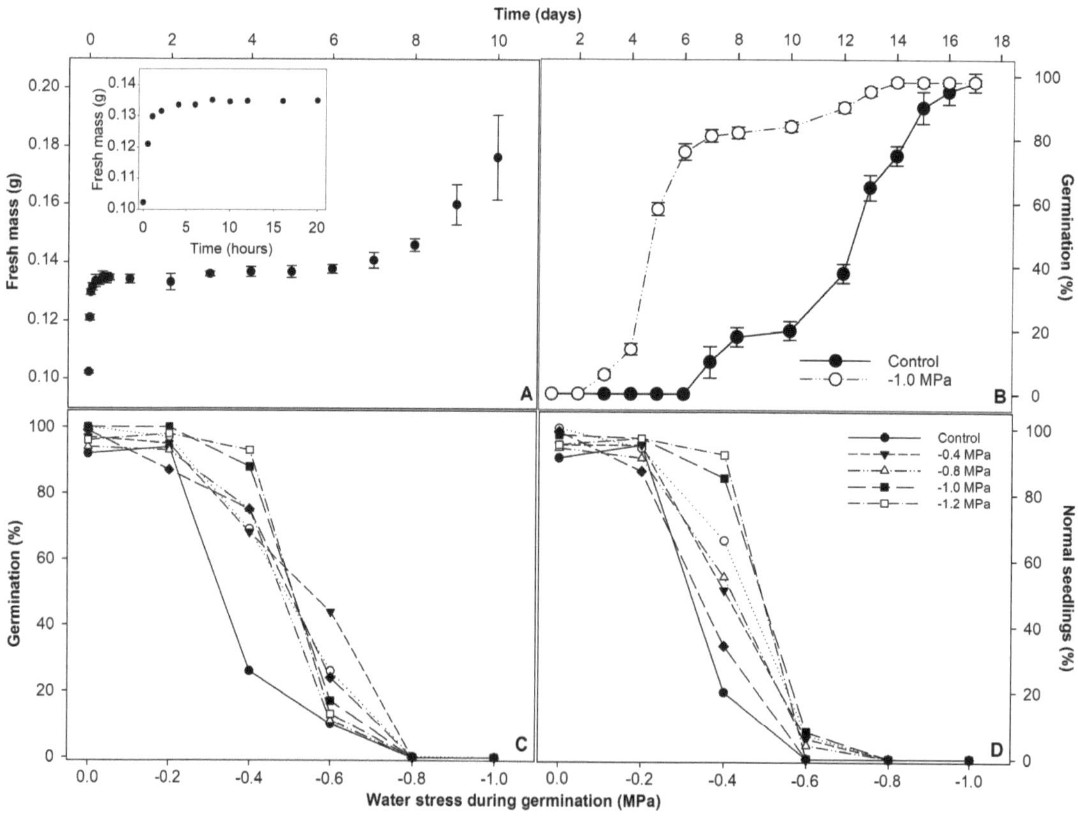

Figure 1. (**A**) Imbibition curve of *Solanum paniculatum* seeds showing phases I, II and III of water uptake. Points are the average of three replicates of 0.1 g each and bars represent standard deviation. The inset shows the faster uptake of water by the dry seeds during phase I. (**B**) Germination of *S. paniculatum* seeds submitted to osmopriming treatment (−1.0 MPa) and the control group (unprimed). (**C**) Germination percentage under water stress of *Solanum paniculatum* control and osmoprimed (−0.4, −0.8, −1.0 and −1.2 MPa) seeds. (**D**) Normal seedling percentage under water stress of *Solanum paniculatum* control and osmoprimed (−0.4, −0.8, −1.0 and −1.2 MPa) seeds. In (**B–D**) symbols represent the average of 4 replicates of 25 seeds each. All seeds were incubated in a germination chamber settled at 25 °C and a 12 h photoperiod.

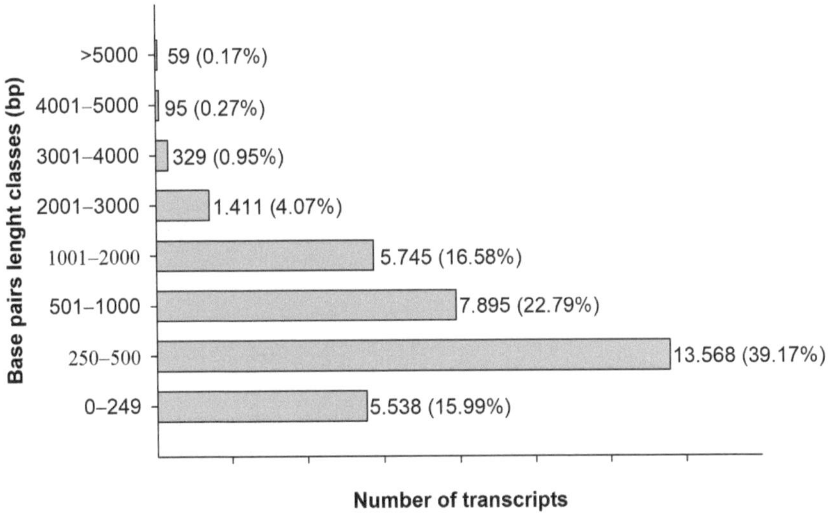

Figure 2. Number and percentages of transcripts in each base pair (bp) length class derived from cDNA libraries of *Solanum paniculatum* control and osmoprimed (−1.0 MPa) seeds. Numbers express the absolute number of base pairs in the transcripts and percentages were generated regarding the whole transcriptome generated with RNAseq analysis.

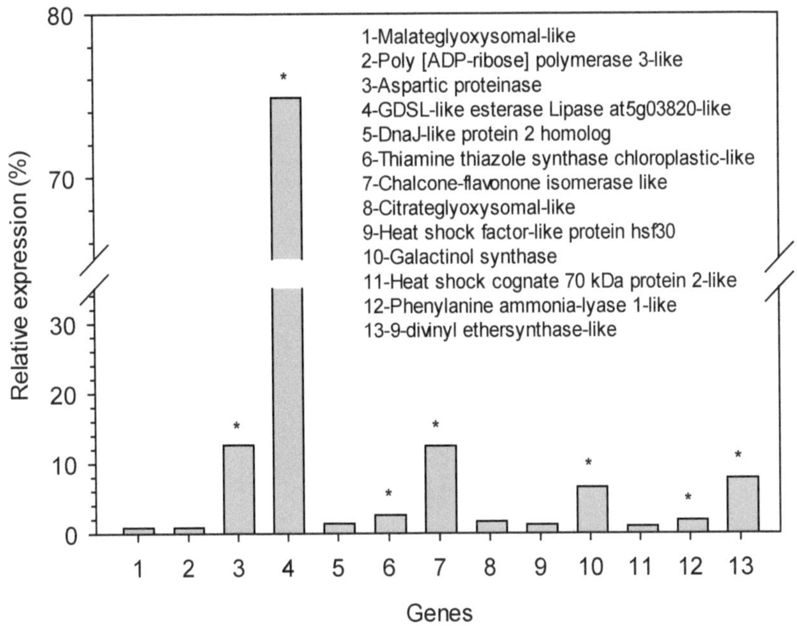

Figure 3. RT-qPCR validation of a specific group of genes that showed differential expression between water-deprived osmoprimed and unprimed *Solanum paniculatum* seeds in RNAseq. * Genes showed a differential expression in osmoprimed seeds ($p < 0.05$).

Table 1. Specific primers and reference genes used in RTq-PCR reactions. The reference genes are indicated by an asterisk.

Contig	Functional Description of Transcript	Size (bp)	Primers Forward (5′-3′)	Primers Reverse (5′-3′)
339	MALATEGLYOXYSOMAL-LIKE	1168	TCCACAACTATGCCAACTTCC	TTTCTCTGCCCTCTCAAACAC
1056	POLY [ADP-RIBOSE] POLYMERASE 3-LIKE	2445	TCCACAACTATGCCAACTTCC	TTTCTCTGCCCTCTCAAACAC
2688	ASPARTIC PROTEINASE	1599	TCAACCGAAACACAAAGGAAG	TTTACCACCGATCAGAACATCA
3417	GDSL-LIKE ESTERASE LIPASE AT5G03820-LIKE	1199	ATGCCTCAACATTGAAGCCT	AGAGCCTTCCCAACAAGATG
3929	DNAJ-LIKE PROTEIN 2 HOMOLOG	1168	ATATTTGTTCCGAGTGCCGA	GTAACATCCCTTTCTCAACTTTCA
7098	THIAMINE THIAZOLE SYNTHASE CHLOROPLASTIC-LIKE	1094	AACCCGTTAAATCAACTCACCA	CGTCATTTCCCTAGCAACAATC
10,620	CHALCONE-FLAVONONE ISOMERASE LIKE	545	AAGAATGAAGTGATGGTGGATGA	CTATGTCTGTTATTCCATGTCCCA
9460	CITRATEGLYOXYSOMAL-LIKE	389	CCAGAGTTTATTGAGGGCGT	CTTCTTCAGCAAGCTTCTTAATCA
6576	HEAT SHOCK FACTOR-LIKE PROTEIN HSF30	595	AGAAAGCAGTATCCACAGCAA	TTAGCCTCAGTATTTCCATCCTC
14,206	GALACTINOL SYNTHASE	637	TCAACTACTCAAAGCTTCGCAT	TATCGCATACACAATCCGCC
8235	HEAT SHOCK COGNATE 70 KDA PROTEIN 2-LIKE	1152	TTCAACTTTCCTCCCAACAG	CAATATCACAGAAATTCGCAGG
6440	PHENYLANINE AMMONIA-LYASE 1-LIKE	1488	GTACAATGCTGTGAAATTCCCT	GAATGGTCAATCATGCTGTCA
21,837	9-DIVINYL ETHERSYNTHASE-LIKE	1115	GGTTACACGACAAATTCATCCC	AGAACACTTTCATGCCTCCAT
* 34	CYTOCHROME P450 87A3-LIKE	1676	TGTATTCTCAAGCTGTCCACT	TTATACCACCTCCAAATGCCA
* 327	HEAT SHOCK PROTEIN 70	989	AGATTACCATCACCAACGACA	GCATAGTTCTCCAAAGCATTCT
* 416	SUBTILISIN-LIKE PROTEASE-LIKE	2737	TGGTGTTGGAGTCGTTGTAG	TGGTGTTGGAGTCGTTGTAG

Table 2. Differentially expressed transcripts (DETs) associated with gene ontology (GO) terms related to response to stress, chemical and abiotic stimulus.

DET (Contig)	Fold-Change (Primed vs. Unprimed)	Functional Description	Gene Ontology Classification		
5548	−53.08	benzoquinone reductase	response to abiotic stimulus	response to chemical	response to stress
488	−26.63	small heat shock protein chloroplastic-like			response to stress
2004	1.52	elongation factor 1-alpha		response to chemical	
1923	1.76	late embryogenesis abundant protein Lea5	response to abiotic stimulus	response to chemical	response to stress
334	1.76	aspartic proteinase-like	response to abiotic stimulus	response to chemical	response to stress
145	1.84	polyadenylate-binding protein 8-like		response to chemical	
194	1.93	dnaJ protein homolog	response to abiotic stimulus		response to stress
9183	2.03	elongation factor 1-alpha		response to chemical	
3485	2.17	peroxidase 12-like			response to stress
4082	2.26	cold shock protein cs66-like	response to abiotic stimulus	response to chemical	response to stress
1334	2.38	cation transport regulator-like protein		response to chemical	
8235	2.48	heat shock cognate 70 kda protein 2-like			response to stress
3929	2.60	dnaj protein homolog 2-like	response to abiotic stimulus		response to stress
16,971	2.92	dehydrogenase/reductase SDR family protein 7-like		response to chemical	
23,208	3.04	heat shock cognate 70 kDa protein 2-like			response to stress
10,417	3.44	heat shock cognate 70 kda protein 2-like			response to stress
2828	3.50	em protein H5-like		response to chemical	response to stress
1719	3.52	non-specific lipid-transfer protein 2-like			response to stress
3594	3.98	17.9 kDa class I heat shock protein-like	response to abiotic stimulus	response to chemical	response to stress
9460	4.32	citrate glyoxysomal-like		response to chemical	response to stress
2973	5.26	11S globulin precursor		response to chemical	
7098	5.32	thiamine thiazole synthase chloroplastic-like	response to abiotic stimulus		response to stress
3565	5.46	tubulin beta-1 chain	response to abiotic stimulus		
7666	5.98	Low-temperature-induced 66	response to abiotic stimulus	response to chemical	response to stress
1638	6.22	11s seed storage globulin		response to chemical	
5501	6.53	RING/U-box domain-containing protein	response to abiotic stimulus	response to chemical	response to stress
6576	7.24	heat shock factor protein hsf30-like	response to abiotic stimulus		response to stress
14,206	7.42	galactinol synthase	response to abiotic stimulus	response to chemical	response to stress
22,593	8.09	chalcone isomerase	response to abiotic stimulus	response to chemical	
10,620	10.51	chalcone isomerase	response to abiotic stimulus	response to chemical	

4. Discussion

Priming is a technique widely used to improve seed quality and to promote faster and more uniform seed germination. In addition, an increased tolerance to abiotic stress is expected and considered one of the main advantages. Analyzing the data presented above, it is noticeable that osmopriming of *Solanum paniculatum* seeds at −1.0 MPa enhanced seed performance under water deficit (Figure 1C,D). The improvement of stress tolerance after priming may be due to a cross-tolerance induced by the osmotic treatment [28]. Apparently, plants have a capacity to "memorize" or develop a "stress imprint" as a genetic or biochemical modification that occurs after stress exposure [13]. The increased capacity of *Solanum paniculatum* seeds to perform better under stress conditions after seed priming may be explained by the enhanced expression of stress-related genes as observed by Song et al. [29], Osthoff et al. [30] and Gao et al. [31] using different stresses as pre-germination treatments.

In fact, by determining the expression profiles of osmoprimed versus unprimed seeds of *S. paniculatum* under water deficiency, we could observe in the present work that a number of genes predicted to code for proteins potentially related to response to stress were differentially expressed between the two conditions (Tables 1 and 2). We discuss below some particular cases.

With A DET predicted to code for a prontein that enables DNA repair, namely, poly [ADP-RIBOSE] polymerase 3-like, displayed clearly enhanced expression upon priming. A similar protein found in *Arabidopsis thaliana* is related to protection against stress caused by gamma radiation [32]. Therefore, it may be argued that this protein plays a role during priming in the protection of DNA and, thus, contributes to successful germination under drought stress. DNA repair genes have been associated with seed vigor and, consequently, seed germination under unfavorable conditions [33].

DETs predicted to code for proteins functionally described as chalcone-flavanone isomerase and galactinol synthase were abundantly transcribed in osmoprimed seeds. Chalcone-flavanone isomerase have been identified in tomato (*Solanum lycopersicum*) and wheat (*Triticum aestivum*), as participating in the biosynthesis of flavonoids by converting chalcone to flavonols [34]. Flavonoids accumulate in plants under stressful conditions and may help the plants to adapt to environmental stress [35]. Galactinol synthase is involved in the synthesis of oligosaccharides from the raffinose family oligosaccharides RFOs [36]. RFOs have various functions in plants; they are used for transport and storage of carbohydrates and act as compatible solutes in the plant cell's protection mechanisms against biotic and abiotic stresses [37]. Our results clearly show that mild stress may induce the transcription of galactinol synthase, a key enzyme associated with the accumulation of galactinol and raffinose under abiotic stress conditions. Apparently, this enzyme functions as an osmoprotectant, favoring tolerance to drought stress during germination of *S. paniculatum*.

DETs predicted to code for proteins belonging to the heat shock proteins family were also found highly expressed upon priming, specially proteins functionally described as shock factor-like protein HSF30, KDA class I heat shock protein 3 and DNAJ-like protein 2 homolog (HSP 40). Heat shock proteins are members of a highly conserved family, known as chaperones, and are constitutively expressed in various organisms, as well as in different cellular compartments [38,39]. These proteins have important functions in plant growth and development, as well as in response to environmental stresses, such as heat and drought. Heat shock proteins act in a variety of cellular processes, including the transport of proteins across membranes, maintenance of proper folding of proteins, regulation of protein degradation and preventing irreversible aggregation of proteins [40]. However, we found a transcript predicted to code for a small heat shock protein chloroplastic-like, which showed a markedly decrease in abundance. Levels of small heat shock proteins (sHSP) have been correlated with longevity in sunflower seeds [41]. Thus, we hypothesize that the downregulation of sHSP proteins during priming may explain why primed seeds generally have reduced longevity [42].

DNAJ-like protein 2 acts similarly to heat shock proteins and is present in plants which are tolerant to salinity [43]. The multifunctional DNAJ-like proteins are encoded by a multigene family and are involved in protein trafficking [44].

5. Conclusions

Osmopriming of *Solanum paniculatum* seeds at −1.0 MPa favors seed germination under water deficit. Proteins related to water, oxidative, saline and heat stresses were upregulated as a result of priming indicating a possible cross-tolerance effect.

Supplementary Materials: The following supporting information can be downloaded at: https://www.mdpi.com/article/10.3390/seeds2040029/s1. Table S1: Final germination percentage during germination under water stress of Solanum paniculatum L. seeds submitted to osmopriming and control treatments; Table S2: Normal seedling percentage during germination under water stress of Solanum paniculatum L. seeds submitted to osmopriming and control treatments; Table S3: Transcripts differentially expressed in primed and unprimed Solanum paniculatum L. seeds. Fold change was calculated using Baggerley's test considering $p < 0.05$; Table S4: Transcripts differentially expressed associated with Solanum paniculatum osmoprimed (−1.0 MPa) seeds.

Author Contributions: Conceptualization, H.W.M.H. and E.A.A.d.S.; data curation, P.B.d.S.; formal analysis, P.B.d.S., T.A.A.V., M.L.A. and L.A.B.; funding acquisition, E.A.A.d.S.; investigation, P.B.d.S., T.A.A.V. and E.A.A.d.S.; methodology, P.B.d.S., T.A.A.V., M.L.A., L.A.B. and E.A.A.d.S.; project administration, E.A.A.d.S.; resources, E.A.A.d.S.; software, L.A.B.; supervision, H.W.M.H. and E.A.A.d.S.; validation, M.L.A. and L.A.B.; writing—original draft, P.B.d.S., T.A.A.V., H.W.M.H. and E.A.A.d.S.; writing—review and editing, T.A.A.V. and E.A.A.d.S. All authors have read and agreed to the published version of the manuscript.

Funding: This work was supported by the São Paulo Research Foundation—FAPESP [grant number #2011/18262-6] and CAPES-WUR program [grant number # 007/09]. PBS thanks Fundação de Amparo à Pesquisa do Estado de São Paulo (FAPESP) for the scholarship and Universidade Estadual Paulista "Julio de Mesquita Filho" (FCA-UNESP/Botucatu) for supporting the experiments. TAAV thanks Fundação de Amparo à Pesquisa do Estado de Minas Gerais (FAPEMIG) for the scholarship and Universidade Federal de Lavras for supporting part of the experiments.

Institutional Review Board Statement: Not applicable.

Informed Consent Statement: Not applicable.

Data Availability Statement: The reads are available in the NCBI Sequence Read Archive, under accession numbers STUDY: PRJNA384240 (SRP105294) SAMPLE: SRX2766227: RNA-seq of *Solanum paniculatum*: unprimed seeds and SAMPLE: SRX2766226: RNA-seq of *Solanum paniculatum*: primed seeds.

Conflicts of Interest: The authors declare no conflict of interest.

References

1. Hussain, H.A.; Hussain, S.; Khaliq, A.; Ashraf, U.; Anjum, A.S.; Men, S.; Wang, L. Chiling and drought stress in crop plants: Implications, cross talk and potential management opportunities. *Front. Plant Sci.* **2018**, *9*, 393. [CrossRef]
2. Vidal, M.C.; Stacciarini-Seraphin, E.; Camara, H.H.L.L. Crescimento de plântulas de *Solanum lycocarpum* St. Hil. (Lobeira) em casa de vegetação. *Acta Botânica Brasílica* **1999**, *13*, 271–274. [CrossRef]
3. Lorenzi, H.; Matos, F.J.A. *Plantas Medicinais No Brasil: Nativas E Exóticas*, 3rd ed.; Instituto Plantarum: Nova Odessa, Brazil, 2021; p. 576.
4. Garcia, J.; Jacobson, T.K.B.; Farias, J.G.; Boaventura, R.F. Efectiveness of methods to increase the germination rate of Jurubeba (*Solanum paniculatum* L.) seeds. *Pesqui. Agropecuária Trop.* **2008**, *38*, 223–226.
5. Martins, A.F. *Contribuição Ao Estudo Das Plantas Medicinais: Manual de Preservação Da Natureza*; Cabral Editora Universitária: São Paulo, Brazil, 1998.
6. Cowan, M.F.; Blomstedt, C.K.; Norton, S.L.; Henry, R.J.; Moller, B.L.; Gleadow, R. Crop wild relatives as a genetic resource for generating low-cyanide, drought-tolerant Sorghum. *Environ. Exp. Bot.* **2020**, *169*, 103884. [CrossRef]
7. Coyne, C.J.; Kumar, S.; Von Wettberg, E.J.B.; Marques, E.; Berger, J.D.; Redden, R.J.; Noel Ellis, T.H.; Brus, J.; Zablastzká, L.; Smykal, P. Potentials and limitis of exploitation of crop wild relatives for pea, lentil and chickpea improvement. *Legume Sci.* **2020**, *2*, e36. [CrossRef]

8. Quezada-Martinez, D.; Nyarko, C.P.A.; Schiessi, S.V.; Mason, A.S. Using wild relatives and related species to build climate resilience in *Brassica* crops. *Theor. Appl. Genet.* **2021**, *134*, 1711–1728. [CrossRef]
9. Kranner, I. What is stress? Concepts, definitions and application in seed science. *New Phytol.* **2010**, *188*, 655–673. [CrossRef]
10. Bowler, C.; Fluhr, R. The role of calcium and activated oxygen as signals for controlling cross-tolerance. *Trends Plant Sci.* **2000**, *5*, 241–246. [CrossRef]
11. Pastori, G.M.; Foyer, C.H. Common components, networks, and pathways of cross-tolerance to stress: The central role of "redox" and abscisic acid-mediated controls. *Plant Physiol.* **2002**, *129*, 460–468. [CrossRef]
12. Capiati, D.A.; País, S.M.; Téllez-Iñón, M.T. Wounding increases salt tolerance in tomato plants: Evidence on the participation of calmodulin-like activities in cross-tolerance signaling. *J. Exp. Bot.* **2006**, *57*, 2391–2400. [CrossRef]
13. Bruce, T.J.A.; Matthes, M.C.; Napier, J.A.; Pickett, J.A. Stressful "memories" of plants: Evidence and possible mechanisms. *Plant Sci.* **2007**, *173*, 603–608. [CrossRef]
14. Balbinot, E.; Lopes, H.M. Efeitos do condicionamento fisiológico e da secagem na germinação e no vigor de sementes de cenoura. *Rev. Bras. De Sementes* **2006**, *1*, 1–8. [CrossRef]
15. Patanèa, C.; Cavallaroa, V.; Cosentino, S.L. Germination and radicle growth in unprimed and primed seeds of sweet sorghum as affected by reduced water potential in NaCl at different temperatures. *Ind. Crops Prod.* **2008**, *30*, 1–8. [CrossRef]
16. Hussain, S.; Hussain, S.; Khaliq, A.; Ali, S.; Khan, I. Physiological, biochemical and molecular aspects of seed priming. In *Priming and Pre-Treatment of Seeds and Seedlings*, 1st ed.; Hasanuzzaman, M., Fotopoulos, V., Eds.; Springer: Singapore, 2019; p. 604.
17. Heydecker, W.; Higgins, J.; Gulliver, R.L. Accelerated germination by osmotic seed treatment. *Nature* **1973**, *246*, 42–44. [CrossRef]
18. Sher, A.; Sarwar, T.; Nawaz, A.; Ijaz, M.; Sattar, A.; Ahmad, S. Methods of Seed Priming. In *Priming and Pre-Treatment of Seeds and Seedlings*, 1st ed.; Hasanuzzaman, M., Fotopoulos, V., Eds.; Springer: Singapore, 2019; p. 604.
19. Adnan, M.; Abd-Ur-Rahman, H.; Asif, M.; Hussain, M.; Bilal, H.M.; Adnan, M.; Ur-Rehman, F.; Ahmad, S.; Khalid, M. Seed priming; An effective way to improve plant growth. *EC Agric.* **2020**, *6*, 1–5.
20. Bradford, K.J. Manipulation of seed water relations via osmotic priming to improve germination under stress conditions. *HortScience* **1986**, *21*, 1105–1112. [CrossRef]
21. ISTA. *International Rules for Seed Testing*; International Seed Testing Association: Zurich, Switzerland, 2004.
22. Michel, B.E. Evaluation of the Water Potentials of Solutions of Polyethylene Glycol 8000. *Plant Physiol.* **1983**, *72*, 66–70. [CrossRef]
23. Villela, F.A.; Beckert, O.P. Potencial osmótico de soluções aquosas de polietileno glicol 8000. *Rev. Bras. De Sementes* **2001**, *23*, 267–275. [CrossRef]
24. Baggerley, K.; Deng, L.; Morris, J.; Aldaz, C. Differential expression in SAGE: Accounting for normal between-library variation. *Bioinformatics* **2003**, *19*, 1477–1483. [CrossRef] [PubMed]
25. Kong, L.; Zhang, Y.; Ye, Z.Q.; Liu, X.O.; Zhao, S.O.; Wei, L.; Gao, G. CPC: Assess the protein-coding potential of transcripts using sequence features and support vector machine. *Nucleic Acids Res.* **2007**, *35*, 345–349. [CrossRef]
26. Conesa, A.; Götz, S.; García-Gómez, J.M.; Terol, J.; Talón, M.; Robles, M. Blast2go: A universal tool for annotation, visualization and analysis in functional genomics research. *Bioinformatics* **2005**, *21*, 3674–3676. [CrossRef] [PubMed]
27. Ramakers, C.; Ruijter, J.M.; Deprez, R.H.; Moorman, A.F. Assumption-free analysis of quantitative real-time polymerase chain reaction (PCR) data. *Neurosci. Lett.* **2003**, *339*, 62–66. [CrossRef] [PubMed]
28. Wahid, A.; Sehar, S.; Perveen, M.; Gelani, S.; Basra, S.M.A.; Farooq, M. Seed pretreatment with hydrogen peroxide improves heat tolerance in maize at germination and seedling growth stages. *Seed Sci. Technol.* **2008**, *36*, 633–645. [CrossRef]
29. Song, L.; Prince, S.; Valliyodan, B.; Joshi, T.; Maldonado Dos Santos, J.V.; Wang, J.; Lin, L.; Wan, J.; Wang, Y.; Xu, D.; et al. Genome-wide transcriptome analysis of soybean primary root under varying water deficit conditions. *BMC Genom.* **2016**, *17*, 57. [CrossRef]
30. Osthoff, A.; Donà Dalle Rose, P.; Baldauf, J.A.; Piepho, H.P.; Hochholdinger, F. Transcriptomic reprogramming of barley seminal roots by combined water deficit and salt stress. *BMC Genom.* **2019**, *20*, 325. [CrossRef]
31. Gao, G.; Hu, J.; Zhang, X.; Zhang, F.; Li, M.; Wu, X. Transcriptome analysis reveals genes expression pattern of seed response to heat stress in *Brassica napus* L. *Oil Crop. Sci.* **2021**, *6*, 87–96. [CrossRef]
32. Song, J.; Keppler, B.D.; Wise, R.R.; Bent, A.F. PARP2 Is the Predominant Poly (ADP-Ribose) Polymerase in Arabidopsis DNA Damage and Immune Responses. *PLoS Genet.* **2015**, *11*, e1005200. [CrossRef] [PubMed]
33. Ventura, L.; Dona, M.; Macovei, A.; Carbonera, D.; Buttafava, A.; Mondoni, A.; Rossi, G.; Balestrazzi, A. Understanding the molecular pathways associated with seed vigor. *Plant Physiol. Biochem.* **2012**, *60*, 196–206. [CrossRef]
34. Winkel-Shirley, B. Flavonoid biosynthesis: A colorful model for genetics, biochemistry, cell biology and biotechnology. *Plant Physiol.* **2001**, *126*, 485–493. [CrossRef]
35. Liu, Z.; Liu, Y.; Pu, Z.; Wang, J.; Zheng, Y.; Li, Y.; Wei, Y. Regulation, evolution, and functionality of flavonoids in cereal crops. *Biotechnol. Lett.* **2013**, *35*, 1765–1780. [CrossRef]
36. Peterbauer, T.; Lahuta, L.B.; Blochl, A.; Mucha, J.; Jones, D.A.; Hedley, C.L.; Gorecki, R.J.; Richter, A. Analysis of the raffinose family oligosaccharide pathway in pea seeds with contrasting carbohydrate composition. *Plant Physiol.* **2001**, *127*, 1764–1772. [CrossRef] [PubMed]
37. Philippe, R.N.; Ralph, S.G.; Bohlmann, S.D.; Mansfield, J. Transcriptome profiles of hybrid poplar (*Populus trichocarpa deltoides*) reveal rapid changes in undamaged, systemic sink leaves after simulated feeding by forest tent caterpillar (*Malacosoma disstria*). *New Phytol.* **2010**, *188*, 787–802. [CrossRef] [PubMed]

38. Wang, W.; Vinocur, B.; Shoseyov, O.; Altman, A. Role of plant heat-shock proteins and molecular chaperones in the abiotic stress response. *Trends Plant Sci.* **2004**, *9*, 244–252. [CrossRef]
39. Waters, E.R. The evolution, function, structure and expression of the plant sHSPs. *J. Exp. Bot.* **2013**, *64*, 391–403. [CrossRef]
40. Su, P.H.; Li, H.M. Stromal Hsp70 is important for protein translocation into pea and Arabidopsis chloroplasts. *Plant Cell* **2010**, *22*, 1516–1531. [CrossRef]
41. Prieto-Dapena, P.; Castaño, R.; Almoguera, C.; Jordano, J. Improved resistance to controlled deterioration in transgenic seeds. *Plant Physiol.* **2006**, *142*, 1102–1112. [CrossRef] [PubMed]
42. Bewley, J.D.; Bradford, K.J.; Hilhorst, H.W.M.; Nonogaki, H. *Seeds: Physiology of Development, Germination and Dormancy*; Springer: New York, NY, USA, 2013; 392p.
43. Zhao, Z.; Zhang, W.; Yan, J.; Zhang, J.; Liu, Z.L.X.; Yi, Y. Over-expression of Arabidopsis DnaJ (Hsp40) contributes to NaCl stress tolerance. *Afr. J. Biotechnol.* **2010**, *9*, 972–978. [CrossRef]
44. Sung, D.Y.; Guy, C.L. Physiological and molecular assessment of altered expression of Hsc70-1 in Arabidopsis. Evidence for pleiotropic consequences. *Plant Physiol.* **2003**, *132*, 979–987. [CrossRef]

seeds

MDPI

Article

Effect of Melatonin on Germination and Seedling Growth in Aging Seeds or under Drought Conditions

Isabel García-Cánovas, Manuela Giraldo-Acosta, Antonio Cano, Marino B. Arnao and Josefa Hernández-Ruiz *

Phytohormones & Plant Development Laboratory, Department of Plant Biology (Plant Physiology), Faculty of Biology, University of Murcia, 30100 Murcia, Spain; isabelmaria.garciac@um.es (I.G.-C.); manuela.giraldoa@um.es (M.G.-A.); aclario@um.es (A.C.); marino@um.es (M.B.A.)
* Correspondence: jhruiz@um.es

Abstract: Seed germination (GS) and seedling growth are vital plant stages that can be affected by stresses such as drought and aging, which cause deterioration and reduce seed viability. With the aim of homogenizing and improving GS, priming treatments with biostimulants such as the antioxidant melatonin are commonly used in seeds. In this study, the effects of melatonin on germination and seedling growth in two different situations, i.e., aging seeds of rice, barley, and sorghum and under polyethylene glycol (PEG)-induced drought stress in sorghum, were studied. Aged seeds were primed for 7 days in different concentrations of melatonin, and drought stress seeds were primed for 24 h before PEG treatment for 7 days, and germination and initial growth parameters were monitored. Aging-seeds of rice and barley showed the maximum response in terms of germination percentage at 20 μM melatonin and 0.05 μM respectively; while aging-seeds of sorghum showed improvement in germination for practically all concentrations studied, even the highest tested at 50 μM. Regarding the effect of melatonin treatments on drought stress in sorghum seeds, all the studied parameters showed a significant attenuation of the adverse effects of drought stress, alleviating them, for all concentrations tested but especially at 200 μM melatonin. The results obtained confirm that priming seeds with melatonin under low germinability conditions relieves stress and improves both germination and seedling growth.

Keywords: melatonin; seeds; germination; abiotic stress; aging; drought; seedling growth

Citation: García-Cánovas, I.; Giraldo-Acosta, M.; Cano, A.; Arnao, M.B.; Hernández-Ruiz, J. Effect of Melatonin on Germination and Seedling Growth in Aging Seeds or under Drought Conditions. *Seeds* **2024**, *3*, 341–356. https://doi.org/10.3390/seeds3030025

Academic Editors: José Antonio Hernández Cortés, Gregorio Barba-Espín and Pedro Diaz-Vivancos

Received: 3 June 2024
Revised: 26 June 2024
Accepted: 1 July 2024
Published: 4 July 2024

1. Introduction

Seeds are the units of sexual reproduction of plants, and their function is to propagate, perpetuate, and disperse the species to which they belong. Seed germination (SG) is a vital stage in plant development and can be considered as a determinant for plant productivity [1]. The major food seeds are those of cereals and legumes; their annual world production is approaching 3 billion tons [2]. SG refers to the physiological process culminating in the emergence of the embryo from its enclosing coverings [3]. It begins with water imbibition, the mobilization of food reserve, protein synthesis, and radicle protrusion. Physiological and biochemical changes followed by morphological changes during germination are strongly related to seedling survival rate and vegetative growth, which affect yield and quality [4].

SG and early seedling growth are two critical stages particularly vulnerable to environmental stress conditions, particularly water stress condition [1]. Drought is a severe limitation for plant growth, development, and productivity, and the response characteristics of plants exposed to drought stress have become a crucial environmental research topic [5]. Drought stress reduces imbibition and increases osmotic potential of growth medium, thus reducing germination percent and seedling vigor [6]. Polyethylene glycol (PEG) with a molecular weight of 6000 Da, is a natural water-soluble and non-ionic polymer [7], which it is generally used to induce drought stress in studies with higher plants [8]. The use of PEG6000 will increase in the osmotic potential of the growth medium, reducing the imbibition capacity by seeds.

Many seeds are capable of surviving dehydration at maturity, and in this state, they can survive for long periods (up to hundreds of years in some cases) and resume growth when rehydrated. However, deteriorative chemical processes continue in dry seeds, resulting in their gradual loss of vigor and eventual death [3]. Viability is a measure of the percentage of seeds with the capacity to germinate and generate seedlings under favorable environmental conditions. The aging of a seed occurs when the viability of a seed in a state of dormancy for long periods of time is affected by deterioration, which will cause a delay in its germination or even its total cessation and/or the reduction in growth seedling rates [9]. Aging involves the impairment of metabolism during germination, with the reduction in the mitochondrial respiratory activity, the production of toxic substances such as oxidized free fatty acids, the reduction in some enzymatic activities, and the loss of soluble compounds due to the abnormal permeability of cell membranes [10].

A good imbibition phase followed by a progressive germination rate, including a rapid and uniform radicle emergence, will improve crop production. It has been observed that the differences in these stages are generally not recovered, directly affecting crop yield [11]. To increase the yield of commercial seed lots, the SG rate can be induced by primming the seeds in different media or immersing them for short times in solutions containing some biostimulant compounds such as phytohormones or others, with the aim of homogenizing and improving SG. Priming technique can promotes quick and uniform germination, improves seedling responses, ensures successful seedling establishment, leading to better plant growth [12]. Seed priming can also enhance the germination of weak, damaged, or aged seeds, even under unfavorable environmental conditions [13].

Melatonin (N-acetyl-5-methoxytryptamine) is an indoleamine derived from tryptophan that was first detected in the bovine pineal gland in 1958 [14]. A few years later, in 1995, it was identified in several plant species using radioimmunoassay and HPLC-MS [15–17]. Melatonin contents in plants are highly variable, ranging from picograms to micrograms per gram of plant material analyzed. Phytomelatonin (plant melatonin) has been detected in a wide variety of plants, from edible to wildtypes, and in all tissues and organs such as seeds, stems, leaves, roots, flowers and fruits [18,19].

This molecule with a molecular weight of 232.2 Da and a chemical formula of C13H16N2O2 is being considered as a new plant hormone, due to its involvement in multiple physiological functions of plant development [20–22]. The identification of the first melatonin receptor (PMTR1/CAND2, a receptor coupled to a G protein) in *Arabidopsis thaliana* was a major breakthrough [23].

At the cellular level, melatonin biosynthesis is located mainly in chloroplasts since they contain the main enzyme responsible for its synthesis, serotonin N-acetyltransferase, which was identified in rice chloroplasts [24] and later in red algae [25]. Other enzymes that participate in the biosynthesis pathway are located in the endoplasmic reticulum, in the cytoplasm and occasionally in the mitochondria [26,27]. In total, five enzymes are involved in its biosynthesis: tryptophan decarboxylase (TDC), tryptamine 5-hydroxylase (T5H), serotonin N-acetyltransferase (SNAT), acetylserotonin methyltransferase (ASMT), and caffeic acid O-methyltransferase (COMT) [28]. These routes are widely studied in Ara-bidopsis thaliana and rice, although some aspects of serotonin biosynthesis still need to be elucidated [29–31]. The biosynthesis route is very similar in both plants and animals, except in certain steps, such as the hydroxylation of tryptophan to 5-hydroxytryptophan by tryptophan hydroxylase (TPH), since this enzyme has not yet been identified in plants, but its substrate and product are present in some plant species [30,32].

Due to the diversity of actions of melatonin described by several authors [20,33,34] either affecting plants in different processes such as germination, rooting, growth, photosynthesis efficiency, parthenocarpy, fruiting, ripening, and senescence, as well as related actions or factors such as stomatal conductance, intercellular CO_2, Rubisco activity, photosystem efficiency, etc., phytomelatonin has been proposed as a master regulator in plants [21].

This denomination is also due to the wide characterization of melatonin as a regulator of gene expression and its relationship with other phytohormones such as auxin, cytokinins, gibberellins, abscisic acid, jasmonic acid, salicylic acid, and ethylene [35,36].

Phytomelatonin contents can be regulated intrinsically (e.g., circadian clock, phytohormones) and extrinsically by biotic and abiotic stressors through several second messengers such as reactive oxygen and nitrogen species (ROS and RNS), gasotransmitter signals (nitric oxide and hydrogen sulfide), and crosstalking with other plant hormones [35,36].

Another characteristic role of melatonin is its function as an antioxidant. This antioxidant capacity has been confirmed in both plant and animal studies. Thus, it has been described that, in plants treated with melatonin and subjected to cadmium stress [37], it indirectly enhances the expression of several antioxidant enzymes, including peroxidases, catalases, and superoxide dismutases [21], and the stress caused by ROS and nitric oxide is regulated [38]. In addition, it has also been seen that melatonin is capable of protecting plant cells against biotic stress caused by pathogens such as fungi, nematodes, bacteria, and viruses [28,39].

Seed priming can improve SG under stress conditions compared to unprimed seeds, as it achieves a rapid and uniform gemination, leading to successful plant establishment. The germination-promoting activity of melatonin was demonstrated in early studies on red cabbage and cucumber [40,41]. In this study, we investigated the optimal concentration of melatonin to mitigate the adverse effect on germination and seedling growth in seeds, under two conditions: aging-seeds (in rice, (*Oryza sativa* L.), barley (*Hordeum vulgare* L.) and sorghum (*Sorghum* spp.)) or drought stress induced by polyethylene glycol (PEG) (in sorghum), determining the adaptative response in the presence of melatonin.

2. Materials and Methods

2.1. Plant and Chemical Materials

Sorghum, rice, and barley seeds used for bioassay I were obtained from a local specialized store in edible seeds (Murcia, Spain) in 2018 and stored at 4 °C for 5 years. For bioassay II, sorghum seeds were purchased from the same local store in 2023.

All seeds used in the different bioassays were classified according to their size and healthy appearance, discarding damaged or small seeds [42].

In all studies, the seeds were sterilized with 10% sodium hypochlorite for 10 min and then washed 3 times with plenty of distilled water to remove traces of the disinfectant.

Melatonin (*N*-acetyl-5-methoxytryptamine) and polyethylene glycol (PEG) were obtained from Sigma-Aldrich (Madrid, Spain). The different melatonin solutions were prepared in distilled water from an initial 10^{-2} M ethanol solution. The drought study was carried out with a PEG6000 solution prepared in distilled water at a concentration of 18% (*w*/*v*), which generated an osmotic potential of $\psi = -0.4$ MPa.

2.2. Study of the Effect of Melatonin Treatment on Aging Seeds

Bioassay I: The study was carried out on the following seeds: rice, barley, and sorghum. A total of 20 seeds previously sterilized were placed per Petri dishes containing two filter paper disks. For priming, 20 mL of distilled water (control) or different concentrations of melatonin (0.05, 0.1, 0.2, 1, 20, 50 µM) was added to each plate (this range of melatonin concentrations was selected by us in previous assays). The plates were placed in a culture chamber with controlled temperature at 25 ± 2 °C and a photoperiod of 16-h light/8-h dark. Each treatment (consisting of 3 petri dishes of 20 seeds each) was repeated 3 times.

2.3. Study of the Effect of Melatonin Treatment on Sorghum Seeds Subjected to Drought Stress

Bioassay II: The drought stress was imposed artificially using polyethylene glycol (PEG6000). Sorghum seeds that were previously sterilized were dipped in different melatonin concentrations, 50, 100, 200, and 300 µM, with distilled water as the control, for 24 h, at 25 ± 2 °C in the dark. After the incubation time, 20 seeds of each treatment were placed in Petri dishes with 2 filter paper discs containing 20 mL of a solution of PEG 18% p/v (−0.4 MPa) [43]. Two types of controls were utilized: seeds not treated with melatonin were sown in water without PEG as control (CK) or sown in PEG 18% (stress control).

The plates were placed in a culture chamber with a controlled temperature of 25 ± 2 °C and a photoperiod of 16 h light/8 h dark. At least three replicates were performed per treatment, and the bioassay was repeated 3 times.

2.4. Determination of Parameters and Indices of Germination

The number of germinated seeds was recorded daily from day 4 to 7 after sowing, for each treatment and repetition. Seed germination was considered to have occurred when the seed coat was broken and a radicle was visible ≥ 0.2 cm. Different indices, germination (G) and germination potential (GP), were calculated according to the formula of [44], relative seed germination (RSG), relative root length (RRL), and germination index (GI) according to the formula of [45], and vigor index (VI) according to [46,47], and by applying following equations:

$$\mathrm{G}(\%) = \frac{\text{Number of germinated seeds}}{\text{Number of seeds kept for germination}} \times 100 \tag{1}$$

$$\mathrm{GP}(\%) = \frac{\text{Number of germinated seeds on the 4rd day}}{\text{Number of seeds kept for germination}} \times 100 \tag{2}$$

$$RSG\ (\%) = \frac{number\ of\ seeds\ germinated\ per\ treatment}{number\ of\ seeds\ germinated\ in\ control} \times 100 \tag{3}$$

$$RRL(\%) = \frac{Mean\ root\ length\ per\ treatment}{Mean\ root\ length\ in\ control} \times 100 \tag{4}$$

$$GI\ (\%) = \frac{GRS \times RRG}{100} \tag{5}$$

$$VI = \frac{seedling\ (radicles + plumules)\ length \times G\ (\%)}{100} \tag{6}$$

2.5. Determination of Seedling Growth

The shoot length of the seedlings was measured, from day 4 to 7 after sowing, for each of the treatments and repetitions. The length was noted from the neck region to the apical end of the leaf. The root length of the seedlings was measured, from day 4 to 7 after sowing, for each of the treatments and repetitions. The length was noted from the base of the stem to the apical end of the main root. On day 7, the dry weight of the seeds that had germinated was determined for each treatment and repetition. For this purpose, seedlings were placed in the oven at 60 °C for 48 h to obtain a constant dry weight, and this weight was recorded on an analytical balance and expressed in gram per plant.

2.6. Stress Tolerance Index (STI)

The stress tolerance index is a useful tool to determine the high yield and stress tolerance potential of seeds assessed in bioassay II (drought stress). Stress tolerance indices for root and shoot growth were estimated on the 7th day. The stress tolerance index in shoot (sSTI) and stress root tolerance index (rSTI) in root were calculated using the following equations [48].

$$sSTI\ (\%) = \frac{\text{Shoot length of stress plant}}{\text{Shoot length of control plant}} \times 100 \tag{7}$$

$$rSTI\ (\%) = \frac{\text{Root length of stress plant}}{\text{Root length of control plant}} \times 100 \tag{8}$$

2.7. Statistical Analyses

The obtained experimental data were statistically analyzed using the IBM SPSS Statistics 22.0 program (International Business Machines Corporation (IBM), New York United States, https://www.ibm.com/support/pages/spss-statistics-220-available-download (accessed on 30 June 2024)). An ANOVA test and a post hoc test with a Tukey "honestly significant difference" (HSD) test were applied for treatment comparisons, with $p < 0.05$, to detect differences between means.

3. Results

3.1. Bioassay I

3.1.1. Germination Study of the Effect of Melatonin Treatment on Aging Seeds

We performed germination bioassays with different concentrations of melatonin (0–50 µM) to determine how it affects the germination of aging seeds. The seeds began to germinate on the third day of incubation at 25 °C, and we recorded the result of the daily germination percentage for each treatment from day 4 to 7. Figure 1 illustrates the germination speed of each treatment for the species tested in bioassay I (rice, barley, and sorghum). The different calculated parameters are shown in Table 1, Table 2, and Table 3, respectively.

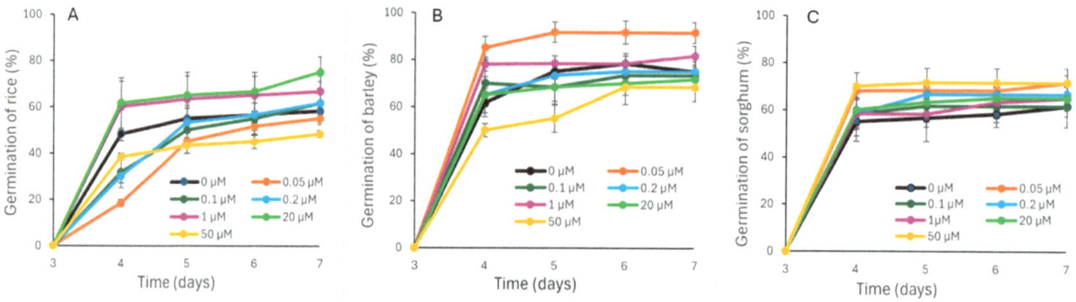

Figure 1. Bioassay I: germination percentage of aging seeds of rice (**A**), barley (**B**), and sorghum (**C**) primed with different concentrations of melatonin from day 4 to day 7. Vertical bars represent data means ± SE ($n = 3$).

Table 1. Effect of different melatonin treatments on the germination and growth of aging rice seeds.

Rice Seeds						
Treatments (µM)	Germination (%)	Germination Potential (%)	Relative Seed Germination (%)	Relative Root Length (%)	Germination Index	Vigor Index
0	58.3 ± 5.0 ^{b,*}	48.3 ± 9.3 ^{c}	100.0 ± 4.1 ^{b}	100.0 ± 21.9 ^{b}	104.7 ± 25.8 ^{b}	1.9 ± 0.5 ^{a}
0.05	55.0 ± 0.1 ^{b}	18.3 ± 1.7 ^{a}	101.6 ± 0.1 ^{bc}	69.3 ± 11.8 ^{a}	70.3 ± 12.0 ^{a}	1.2 ± 0.2 ^{a}
0.1	61.7 ± 9.3 ^{c}	31.7 ± 6.7 ^{b}	113.9 ± 17.1 ^{c}	69.7 ± 12.0 ^{a}	83.4 ± 27.0 ^{bc}	1.4 ± 0.4 ^{a}
0.2	61.7 ± 6.0 ^{c}	30.0 ± 2.9 ^{b}	114.6 ± 11.3 ^{c}	70.6 ± 12.5 ^{a}	81.7 ± 18.5 ^{bc}	1.4 ± 0.2 ^{a}
1	66.7 ± 4.4 ^{cd}	60.0 ± 2.5 ^{d}	111.4 ± 8.6 ^{bc}	154.7 ± 10.2 ^{c}	172.5 ± 18.1 ^{c}	3.1 ± 0.4 ^{b}
20	68.3 ± 6.7 ^{d}	61.7 ± 2.5 ^{d}	114 ± 14.3 ^{c}	155.8 ± 9.1 ^{c}	180.4 ± 30.9 ^{c}	3.8 ± 0.3 ^{b}
50	48.3 ± 1.7 ^{a}	38.3 ± 1.7 ^{b}	77.1 ± 4.9 ^{a}	88.0 ± 4.5 ^{b}	68.3 ± 7.4 ^{a}	1.5 ± 0.1 ^{a}

* Values are mean ± standard error (SE, $n = 3$). Different letters indicate significant differences using the Tukey HSD test ($p < 0.05$).

Table 2. Effect of different melatonin treatments on the germination and growth of aging barley seeds.

			Barley			
Treatments (μM)	**Germination (%)**	**Germination Potential (%)**	**Relative Seed Germination (%)**	**Relative Root Length (%)**	**Germination Index**	**Vigor Index**
0	75.0 ± 0.1 [b,*]	61.7 ± 6.0 [b]	100.0 ± 0.1 [b]	100.0 ± 4.3 [d]	100.0 ± 4.3 [d]	5.2 ± 0.1 [c]
0.05	91.7 ± 4.4 [d]	85.0 ± 5.0 [d]	122.2 ± 5.9 [d]	82.7 ± 2.4 [c]	101.0 ± 4.8 [d]	6.1 ± 0.3 [d]
0.1	73.3 ± 6.0 [ab]	70.0 ± 8.7 [c]	97.8 ± 8.0 [ab]	87.2 ± 0.5 [d]	85.2 ± 6.4 [c]	4.9 ± 0.3 [c]
0.2	75.0 ± 2.9 [b]	65.0 ± 5.0 [bc]	100.0 ± 3.8 [b]	90.0 ± 0.9 [d]	90.1 ± 4.4 [c]	5.0 ± 0.2 [c]
1	81.7 ± 4.4 [c]	80.0 ± 2.5 [d]	108.9 ± 5.9 [c]	79.4 ± 1.0 [c]	86.4 ± 4.3 [c]	4.8 ± 0.4 [c]
20	71.7 ± 4.4 [ab]	65.0 ± 12.5 [bc]	96.7 ± 10.0 [ab]	73.3 ± 4.8 [b]	70.0 ± 5.7 [b]	4.0 ± 0.5 [b]
50	68.3 ± 6.0 [a]	51.7 ± 3.3 [a]	91.1 ± 8.0 [a]	54.0 ± 3.4 [a]	49.3 ± 5.7 [a]	2.5 ± 0.3 [a]

* Values are mean ± standard error (SE, n = 3). Different letters indicate significant differences using the Tukey HSD test ($p < 0.05$).

Table 3. Effect of different melatonin treatments on the germination and growth of aging sorghum seeds.

			Sorghum			
Treatments (μM)	**Germination (%)**	**Germination Potential (%)**	**Relative Seed Germination (%)**	**Relative Root Length (%)**	**Germination Index**	**Vigor Index**
0	61.7 ± 1.7 [a,*]	62.5 ± 2.5 [a]	100.0 ± 3.6 [a]	100.0 ± 11.1 [a]	100.5 ± 14.2 [a]	2.0 ± 0.1 [a]
0.05	71.7 ± 3.3 [b]	70.0 ± 2.9 [b]	116.2 ± 5.4 [b]	118.7 ± 15.2 [a]	139.3 ± 22.7 [b]	2.7 ± 0.5 [b]
0.1	61.7 ± 8.8 [a]	58.3 ± 9.3 [a]	100.0 ± 14.3 [a]	181.5 ± 8.0 [b]	179.4 ± 19.3 [c]	3.1 ± 0.3 [bc]
0.2	66.7 ± 3.3 [ab]	58.3 ± 6.7 [a]	108.1 ± 5.4 [ab]	186.4 ± 4.3 [b]	201.1 ± 6.2 [c]	3.4 ± 0.2 [c]
1	65.0 ± 7.6 [ab]	58.3 ± 11.7 [a]	105.4 ± 12.4 [ab]	178.9 ± 38.9 [b]	197.2 ± 62.7 [c]	4.1 ± 0.9 [d]
20	65.0 ± 5.0 [ab]	60.0 ± 12.5 [a]	101.4 ± 12.2 [ab]	103.9 ± 0.3 [a]	109.6 ± 8.6 [ab]	2.5 ± 0.4 [ab]
50	71.7 ± 6.0 [b]	70.0 ± 5.8 [b]	116.2 ± 9.7 [b]	105.1 ± 8.0 [a]	123.5 ± 18.1 [ab]	2.8 ± 0.4 [b]

* Values are mean ± standard error (SE, n = 3). Different letters indicate significant differences using the Tukey HSD test ($p < 0.05$).

The evolution of the germination percentage from the beginning of data collection, sampling from day 4 to day 7 from sowing is shown in Figure 1. The final germination data of day 7 for each tested seed type (rice, barley, and sorghum) are shown in Table 1, Table 2, and Table 3, respectively. In general, it is observed that the maximum germination rate occurs on day 4 for all treatments and seeds, and from this day, the germination rate slows down and remains the same. This behavior is especially appreciated in sorghum, while in rice and barley, some variation is observed between days 4 and 5.

In aging rice seeds, treatments containing 0.1, 0.2, 1, and 20 μM of melatonin induced an increase in germination percentage (G) compared to the control, with significant differences according to Tukey's post hoc test ($p < 0.05$) (Table 1). The 20 μM melatonin treatment was the most effective, with an increase in the germination percentage of 10.0% compared to the control (Table 1). It should be noted that the maximum tested concentration of 50 μM melatonin had a deleterious effect, with a G reduction of 10.0% compared to the control, and this behavior is also observed in the values obtained for GI and RSG, i.e., the parameters that inform about the behavior of the treated seeds in relation to the control (Table 1). According to Table 1, 20 μM melatonin treatment was, of all the tested concentrations, the one that shows an improved value in all the calculated parameters compared to the control; thus, it can be concluded that, in aging rice seeds, treatment with 20 μM melatonin represents a significant improvement in germination.

In aging barley seeds, 0.05 and 1 μM melatonin treatments caused a significant increase in the germination percentage (G) compared to control (Figure 1), but of these two, 0.05 μM melatonin was clearly more effective, with an increase compared to the control for G of 16.74% compared to the increase of 6.7% caused by 1 μM melatonin (Table 2). Furthermore, 0.05 μM melatonin shows a significant improvement, not showed by melatonin 1 μM, for most of the calculated parameters, with 0.05 μM melatonin being the only one that improve the vigor index (VI) with respect to the control. In barley seeds, as in rice seeds, the maximum tested concentration (50 μM) most affects germination (G) compared to the control, reducing it to 68.3 (−6.7%). This effect is even more deleterious than in rice, since all the calculated parameters are affected and exhibit decreased values for all the tested concentrations. It can be concluded that the concentration of melatonin capable of improving germination in barley is the lowest tested concentration (0.05 μM).

In sorghum, as shown in Table 3, and unlike what happens with the other two species of studied aging seeds (rice and barley, Tables 1 and 2, respectively), at all tested melatonin concentrations (except 0.1 μM) and for calculated parameters, an improvement is observed with respect to the control, resulting in higher (0.05 and 50 μM melatonin) or at least equal values (without significant difference with the control, 0.2, 1, and 20 μM) for G, GP, SRG RRL, IG, and VI. It should be noted that the concentration of melatonin with the highest VI (parameter that combines the values of the germination percentage with the growth of the seedling) is 1 μM followed by 0.2 μM, which induce a VI that is 2.1 and 1.7 times higher the control, respectively. It can be concluded that, in sorghum seeds, the melatonin concentration capable of further improving germination was the lowest tested concentration (0.05 μM). Notably, the highest tested melatonin concentration (50 μM) reduces G for both aging rice and barley, unlike the response of aging sorghum where G clearly improves with 50 μM melatonin.

3.1.2. Effect of Melatonin Treatment on Seedling Growth

In rice, in relation to shoot and root growth (Figure 2A,D), the results revealed statistically significant improvements in 1 and 20 μM melatonin treatments compared to the control. Rice seeds treated with 1 μM melatonin showed the greatest effect in relation to stem elongation, reaching a value of 1.86 cm, 45.31% higher than that in the control with a value of 1.28 cm (Figure 2A). For root elongation, 20 μM melatonin treatment showed the greatest effect, reaching a value of 3.13 cm, 51.94% higher than that for the control (Figure 2D).

In barley, 0.05 and 0.1 μM melatonin treatments induced shoot length with respect to the control. Among these treatments, 0.05 μM melatonin showed the greatest increase in shoot length, reaching a value of 3.86 cm, which represents an increase of 10% compared to the control. A 0.2–50 μM melatonin interval provoked some decrease in shoot growth, which increases as the tested concentration of melatonin increases (Figure 2B). In relation to the root growth of barley, a decrease in root length is observed at all tested concentrations of melatonin, with this effect being greater in the highest concentrations, i.e., 20 and 50 μM melatonin (Figure 2E).

In sorghum, a significant increase in shoot growth compared to the control was observed for all tested melatonin treatments except for the 0.05 μM melatonin (Figure 2C). The maximum shoot length-promoting effect was observed at 1 μM melatonin, with a height of 2.59 cm, 112% higher than that of the control; higher melatonin treatments provoked lower length responses in sorghum shoots (Figure 2C). In relation to the roots, a significant increase in root length was observed compared to the control for the 0.1, 0.2, and 1 μM melatonin treatments, showing a typical Gaussian bell response, with a maximum response at 0.1 μM melatonin, with a value of 4.36 cm, which represents an increase of 122% compared to the control (Figure 2F).

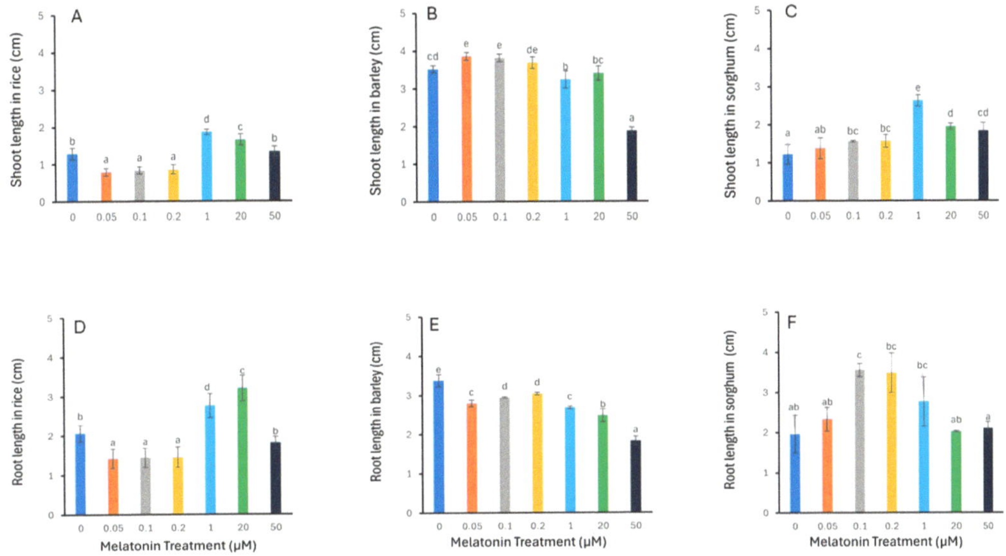

Figure 2. Bioassay I: effect of melatonin treatment on shoot and root growth in aging seeds of rice (**A**,**D**), barley (**B**,**E**), and sorghum (**C**,**F**) at 7 days. Different letters indicate significant differences based on a Tukey HSD test with a significance level of $p < 0.05$. Vertical bars represent data means ± SE ($n = 3$).

In relation to the effect of melatonin treatments on dry weight seedlings (Figure 3), a similar behavior is observed in rice and sorghum, with a very notable increased value at 0.2 μM melatonin compared to the control. Thus, the dry weight of sorghum seedlings increased by approximately 4-fold while that of rice seedlings increased by 1.5-fold compared to the control. On the other hand, barley does not show improvement with any of the tested melatonin treatment.

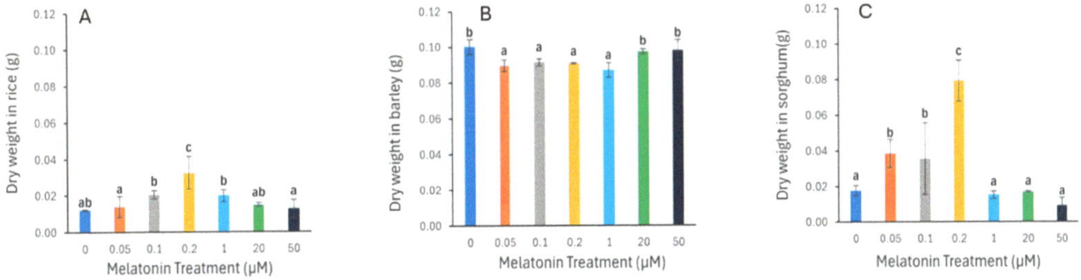

Figure 3. Bioassay I: dry weight of seedlings from melatonin-treated aging seeds of rice (**A**), barley (**B**), and sorghum (**C**) at day 7 of sowing. Different letters indicate significant differences based on a Tukey HSD test with a significance level of $p < 0.05$. Vertical bars represent data means ± SE ($n = 3$).

3.2. Bioassay II

3.2.1. Germination Study of the Effect of Melatonin in Drought Stress Conditions

Sorghum seeds (non-aging) subjected to drought stress induced by 18% PEG show a decrease in the germination percentage of 15.3% compared to the control (sown in water), which denotes the effectiveness of PEG to provoke drought stress in seeds (Table 4). Regarding the studied germination parameters in melatonin-primed seeds in the presence of PEG, a general improvement in their germination can be observed for all tested melatonin concentrations (50, 100, 200, and 300 μM). It should be noted that 200 μM melatonin is even capable of increasing germination (G) by 8.4% more than that in the control. The behavior observed with the different concentrations of melatonin is similar to the Gaussian bell that is observed in the classic growth test of coleoptiles incubated with the auxin indole 3-acetic acid, where both low and high concentrations cause a weak effect either due to either lack or excess of phytohormone (probably due to saturation that causes toxicity). This same behavior was observed for all calculated parameters GP, RSG, GI, and VI. It should be noted that the VI has a similar value for 100 and 200 μM melatonin treatments, being approximately 2.5-times higher in the PEG control, not exceeding in any case the value of the control (CK), although as it has been mentioned above, the germination induced by 200 μM melatonin is greater than that of CK. This is because VI is a parameter that encompasses germination and plant growth, and the growth under drought stress conditions is affected by the low availability of water, which plays an important role in cell elongation. Thus, PEG-induced drought stress effectively inhibits sorghum seed germination, with 200 μM melatonin being an improvement treatment alleviating the drought stress.

Table 4. Germination parameters in sorghum drought stress seeds primed with different melatonin concentrations.

			Sorghum + PEG 18%			
Treatments (μM)	**Germination (%)**	**Germination Potential (%)**	**Relative Seed Germination (%)**	**Relative Root Length (%)**	**Germination Index**	**Vigor Index**
CK	68.3 ± 10.1 [bc,*]	68.3 ± 10.1 [bc]	100.0 ± 14.8 [bc]	100.2 ± 2.8 [e]	100.2 ± 15.5 [e]	1.7 ± 0.4 [d]
P	53.3 ± 10.9 [a]	53.3 ± 10.9 [a]	77.9 ± 16.0 [a]	37.6 ± 3.9 [a]	28.1 ± 2.9 [a]	0.4 ± 0.1 [a]
PM50	66.7 ± 4.4 [b]	66.7 ± 4.4 [b]	97.3 ± 6.4 [bc]	61.0 ± 5.7 [c]	58.7 ± 2.0 [c]	0.8 ± 0.1 [bc]
PM100	73.3 ± 6.0 [bc]	73.3 ± 6.0 [bc]	104.1 ± 10.0 [bc]	61.3 ± 6.2 [c]	65.0 ± 13.0 [c]	0.9 ± 0.2 [c]
PM200	76.7 ± 6.0 [c]	76.7 ± 7.5 [c]	111.9 ± 8.8 [c]	70.7 ± 4.2 [d]	79.1 ± 8.2 [d]	1.0 ± 0.1 [c]
PM300	65.0 ± 2.9 [b]	65.0 ± 2.9 [b]	94.9 ± 7.3 [b]	48.4 ± 6.1 [b]	45.6 ± 5.3 [b]	0.5 ± 0.1 [ab]

* Values are mean ± standard error (SE, $n = 3$). Different letters indicate significant differences using the Tukey HSD test ($p < 0.05$). CK, control absolute; P, PEG stress without melatonin; PM50, PEG stress with 50 μM melatonin; PM100, PEG stress with 100 μM melatonin; PM200, PEG stress with 200 μM melatonin; and PM300, PEG stress with 300 μM melatonin.

3.2.2. Melatonin Effect on Sorghum Seedling Growth in Drought Stress Conditions

Sorghum shoot length decreased significantly in the presence of PEG compared to the control seedlings (Figure 4A). A slight improvement in the shoot growth of the stressed seeds is observed for all the melatonin treatments, except for the one with the highest concentration (300 μM melatonin). In the case of the effect on root growth, PEG-stressed seeds showed a drastic decrease (more than double) than control seeds (without PEG), with values of 0.53 cm (P0 in Figure 4B) and 1.42 cm (CK in Figure 4B), respectively. As in sorghum shoots, a general improvement was observed when seeds were primed with melatonin; the melatonin-induced effect on root growth was stronger, showing all PEG-stressed seeds primed with melatonin (50, 100, 200, and 300 μM), an increased growth response compared to the PEG control.

The results of shoot and root growth were reflected in the dry weight values of seedling as shown in Figure 4C, where it is observed that the lowest dry weight value is obtained for the control with PEG, with better values for the seedlings treated with melatonin. In any case, shoot, root, and dry weight reached the respective values presented by the non-stressed seeds (Figure 4).

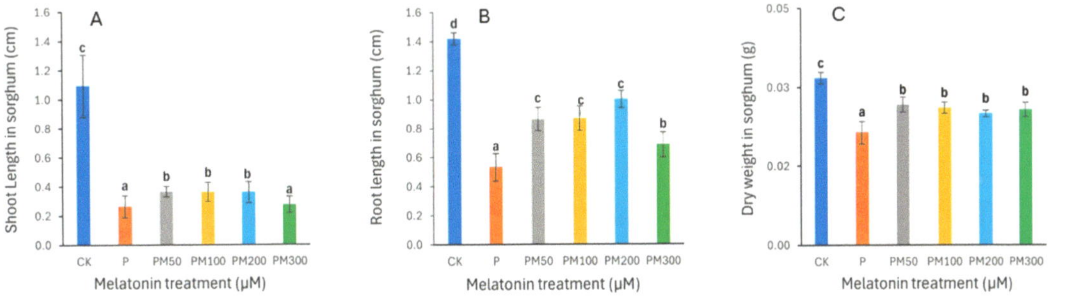

Figure 4. Bioassay II: effects of melatonin treatment on the seedling growth of sorghum under drought stress induced by PEG 18%: (**A**) shoot length, (**B**) root length, and (**C**) dry weight. CK, control; P, PEG stress without melatonin; PM50, PEG stress with 50 μM melatonin; PM100, PEG 18% stress with 100 μM melatonin; PM200, PEG stress with 200 μM melatonin; and PM300, PEG stress with 300 μM melatonin. Different letters indicate significant differences based on a Tukey HSD test with a significance level of $p < 0.05$. Vertical bars represent data means ± SE ($n = 3$).

3.2.3. Stress Tolerance Index (STI) in Melatonin-Primed Sorghum Seeds

The tolerance to drought stress of seeds treated with melatonin was improved compared to the untreated seeds in both roots and shoots of seedlings (Table 5). It was observed that stress tolerance increased by 13% for 50, 100, and 200 μM of melatonin for the shoots. The effect of increasing tolerance is stronger in the roots than that in the shoots, and the highest improvement was observed at 200 μM melatonin, presenting an increase in tolerance of 33.1% compared to the control. The rest of the applied melatonin concentrations also improve tolerance, with the highest used concentration of melatonin (300 μM) having a minor effect.

Table 5. Results of PEG-induced drought stress tolerance indices for primed sorghum seeds with different melatonin concentrations.

	Sorghum +PEG 18%	
Treatments (μM)	**Stress Tolerance Index in Shoot (%)**	**Stress Tolerance Index in Root (%)**
CK	100.0 ± 19.0 ^{b,*}	100.0 ± 2.9 ^{e}
P	20.6 ± 3.2 ^{a}	37.6 ± 3.9 ^{a}
PM50	33.6 ± 3.4 ^{a}	61.0 ± 6.2 ^{c}
PM100	33.0 ± 5.5 ^{a}	61.3 ± 5.7 ^{c}
PM200	32.7 ± 6.7 ^{a}	70.7 ± 4.2 ^{d}
PM300	24.2 ± 4.0 ^{a}	48.4 ± 6.1 ^{b}

* Values are mean ± standard error (SE, $n = 3$). Different letters indicate significant differences using the Tukey HSD test ($p < 0.05$). CK, control; P, PEG stress without melatonin; PM50, PEG stress with 50 μM melatonin; PM100, PEG stress with 100 μM melatonin; PM200, PEG stress with 200 μM melatonin; and PM300, PEG stress with 300 μM melatonin.

In summary, treatments with melatonin through the priming of seeds help to attenuate the effect of drought stress on seedlings, with roots being the organ that responds best to the presence of melatonin.

4. Discussion

The results obtained in this study show that melatonin is a promising agent for improving SG and growth of seedlings, under both studied conditions, i.e., aging seeds and drought-stressed seeds. In general, the effect of melatonin seem to be conditioned by the concentration used [49,50], as it has been demonstrated in previous studies that high concentrations of melatonin inhibited or did not exert a stimulating effect on germination, while low concentrations exerted a stimulating effect [51–55]. Here, a similar behavior is observed with aging seeds of barley and rice, where the maximum effect on the germination percentage is obtained with a melatonin concentration of 0.05 μM in barley and 20 μM in rice (16.7 and 10% more germination than the control, respectively), and it was inhibited in both species at 50 μM melatonin, the highest tested concentration (6.7 and 10% less germination than control, respectively). While aging seeds of sorghum show an improvement in germination for practically all studied concentrations, even the highest tested concentration of 50 μM.

In relation to the effect of melatonin on the elongation of shoots and roots, it was observed that the response of both organs was different and dependent on the concentration of melatonin used, influencing the sensitivity of the organ. This behavior is already well known in the case of auxin, where roots are much more sensitive to auxin than stems or leaves. The influence of melatonin on plant growth was first described by Arnao and colleagues [50,56]. Our study found that melatonin, similar to IAA, promoted vegetative growth in the etiolated hypocotyls of *Lupinus albus* L. and the etiolated coleoptiles of some monocotyledons (canary seed, wheat, barley, and oat) and significantly inhibited root growth. In the current study, melatonin also inhibited the root growth of barley compared to that of the control. However, in sorghum, melatonin promoted the growth of both roots and shoots, with roots showing greater sensitivity compared to shoots, which requires a ten-fold higher concentration. In rice seedlings, melatonin promoted both root and shoot growth at similar concentrations.

In other studies of wild mustard seedlings [57], at low melatonin concentrations, the biosynthesis of IAA was stimulated, and, on the contrary, when melatonin concentration was increased, then, it exerted an inhibitory effect on IAA. This result coincides with that of other recent studies, in which the cause of this behavior is revealed: the higher concentrations of auxin promote cell expansion in shoots but inhibit cell expansion in roots. Auxin is involved in the roots in the activation of two signaling pathways that act antagonistically, i.e., (1) a transmembrane kinase1 (TMK1) pathway and (2) the transport inhibitor response1 and auxin-signaling f-box (TIR1/AFB) pathway, where the first is based on the acidification of the apoplast, which facilitates cell expansion, and the second on its alkalization [58]. Furthermore, the growth mechanisms in melatonin-treated seedlings remain uncertain. In previous studies, the ability of melatonin to positively affect the development of the shoots and radicles was studied by the reduction in intercellular pH and the relaxation of the cell wall induced by melatonin, as could be inferred from the elongation and expansion of the cell wall in lupin [59].

Therefore, there is a relationship between IAA and melatonin, which is associated with the fact that both hormones use the same precursor, L-tryptophan. Furthermore, both compounds mutually modulate their contents: the low levels of exogenous IAA enhance melatonin production, while the high levels of melatonin decrease IAA production and also reduce the levels of auxin-transporting PIN proteins [57,60]. In general, it is observed that there is an interaction between melatonin and IAA in their transcriptional regulation, both upstream and downstream. All these interactions seen together would modulate the development of plants and their adaptation to stress [61].

Aging affects seeds, causing a loss of germinability. The ability to retain germinability, vigor, and viability varies widely among plant species [62]. Aging effect has been studied in different seeds such as *Arabidopsis* and common beech [63,64], and in general, it was observed that seed aging reduces germination by increasing ROS production (superoxide anion radicals, hydrogen peroxide, and hydroxyl radicals), lipid peroxidation membrane injury, and DNA alterations [65]. Therefore, the antioxidant capacity of melatonin would justify the results obtained pertaining to the increase in germination in aging rice, barley, and sorghum seeds assessed in this study. A few studies have been published that have assessed the participation of melatonin in the viability or aging of seeds during long-term storage. It is worth highlighting the studies carried out on pepper [66] and lettuce [67] consisted of pretreating the seeds in melatonin solutions for 24 h and then drying them and storing them for 1 year for pepper or 2 years for lettuce. Pepper seeds pretreated with 25 μM melatonin resulted in a higher germination percentage and lower germination rate and electrical conductivity values during storage. In lettuce seeds, it was observed that the quality of non-pretreated seeds deteriorated rapidly when stored at 25 °C, but pretreatment with melatonin significantly reduced this deterioration by increasing the activities of antioxidant enzymes and restoring membrane properties. The authors conclude that melatonin could be used to slow the natural aging process of seeds and may also have important practical applications, especially in storing seeds of endangered species or valuable breeding material.

About the effect of drought stress induced by PEG in sorghum seeds (bioassay II), this stress could inhibit their germination and plant growth. A significant decrease in the germination percentage, vigor index, germination index, and shoot and root growth of sorghum seeds sown at −0.4 MPa was observed. However, after melatonin treatment, all the above studied indicators (G, GP, RSG, RRL, GI, and VI) showed a significant attenuation of adverse effects of drought stress, alleviating them, which was confirmed when the tolerance stress indices of the shoot and root were calculated (Table 5), where tolerance increased with the majority of tested melatonin concentrations (except for the highest concentration used, 300 μM). These results are consistent with the previous studies carried out in cucumber [68], rice [44], and carrot. In carrot, the effect of osmopriming melatonin on SG was closely related to melatonin treatment, being the optimal melatonin concentration capable of promoting SG between 50 and 200 μM [69]. This behavior can be explained by the antioxidant capacity of melatonin that can reduce the amount of ROS that is produced in the presence of PEG. Studies carried out on seed soybean stressed with PEG showed an excess of intracellular ROS, capable of damaging the plant membranes, thereby causing lipid peroxidation. Melatonin treatment reduced this effect, by decreasing H_2O_2 content and O_2^- production [70]. Similar results have been observed in a study on triticale seeds (*Triticale hexaploide* L.), where priming seeds with 20 μM melatonin alleviated the adverse effects of drought stress on germination and seedling growth induced by PEG6000. The results suggested that the priming of seeds with melatonin promotes ROS-scavenging capacity and improves energy supply and antioxidant enzymatic activities, to alleviate the adverse effects of drought stress on triticale [71].

Melatonin improve germination and seedling growth through the modulation of plant hormone contents in tissues, mainly on auxin, gibberellins, cytokinins, abscisic acid, ethylene, jasmonic acid, salicylic acid, and brassinosteroids [20]. In aging or stressful conditions, SG is clearly affected. To achieve a sustainable level of crop yield, it is important to improve SG under abiotic stress conditions. A recent review has been published on the effects of stressors on SG and the regulatory role that melatonin can play due to its ability to interact with different physiological mechanisms [65]. This review highlights that melatonin induces specific responses to stress, such as the regulation of ionic homeostasis and hydrolysis of storage proteins under salinity stress, the activation of aquaporins, and the accumulation of osmolytes under drought stress. Melatonin also modulates common responses such as its role in the regulation of gibberellin biosynthesis, abscisic acid catabolism, redox homeostasis, and Ca^{2+} signaling, all of which are important players in germination.

5. Conclusions

- The exogenous application of melatonin in aging seeds has a biostimulator effect; this effect depends on the seed under study and the concentration of melatonin applied.
- As occurs in other cases applying phytoregulators, the effectiveness of melatonin depends on the tissue sensitivity, observing induction or inhibition of growing in different tissues (stem and roots) at similar melatonin concentrations.
- In relation to the aged seeds of rice, barley and sorghum tested, the most effective melatonin concentration, especially in the germination parameters and vigor index, has been 20, 0.05 and 1 μM respectively.

- PEG-induced drought stress in sorghum is alleviated by almost all melatonin concentrations (50, 100 and 200 μM) tested, but the 200 μM concentration stands out for being the most effective in improving parameters of germination, early growth and stress.

To conclude, the data obtained in this study focus on the biostimulary role of melatonin in the germination process, reinforcing seedlings against abiotic stressors, through the modulation of redox network and plant hormonal responses.

Author Contributions: Conceptualization, M.B.A. and J.H.-R.; methodology, J.H.-R., M.G.-A. and A.C.; software, M.G.-A. and I.G.-C.; validation, M.B.A. and J.H.-R.; formal analysis, A.C.; investigation, I.G.-C.; resources, I.G.-C.; data curation, I.G.-C.; writing—original draft preparation, I.G.-C.; writing—review and editing, J.H.-R.; visualization, A.C.; supervision, J.H.-R., M.B.A. and A.C.; project administration, J.H.-R. and M.B.A.; funding acquisition, M.B.A. All authors have read and agreed to the published version of the manuscript.

Funding: This research was funded by the Ministerio de Ciencia e Innovación (Spain), Proyectos de I+D+i, Programas Estatales de Generación de Conocimiento y Fortalecimiento Científico y Tecnológico del Sistema de I+D+i y de I+D+i Orientada a los Retos de la Sociedad, del Plan Estatal de Investigación Científica y Técnica y de Innovación 2017–2020, Grant PID2020-113029RB-I00, by MCIN/AEI/10.13039/501100011033. This study forms part of the Agroalnext programme and was supported by MCIU with funding from European Union NextGenerationEU (PRTR-C17.I1) and by Comunidad Autónoma de la Región de Murcia—Fundación Séneca. Giraldo-Acosta, M. has a Predoctoral Contract FPUUM-2023 from the University of Murcia (Spain). More information can be found at Phytohormones and Plant Development Research Lab: https://www.um.es/en/web/phytohormones.

Institutional Review Board Statement: Not applicable.

Informed Consent Statement: Not applicable.

Data Availability Statement: The data presented in this study are available on request from the corresponding author.

Conflicts of Interest: The authors declare no conflicts of interest.

References

1. Ali, A.S.; Elozeiri, A.A.; Ali, A.S.; Elozeiri, A.A. Metabolic Processes During Seed Germination. In *Advances in Seed Biology*; IntechOpen: London, UK, 2017; ISBN 978-953-51-3622-4.
2. Bewley, J.D.; Bradford, K.J.; Hilhorst, H.W.M.; Nonogaki, H. Structure and Composition. In *Seeds: Physiology of Development, Germination and Dormancy*, 3rd ed.; Bewley, J.D., Bradford, K.J., Hilhorst, H.W.M., Nonogaki, H., Eds.; Springer: New York, NY, USA, 2013; pp. 1–25, ISBN 978-1-4614-4693-4.
3. Bewley, J.D.; Bradford, K.J.; Hilhorst, H.W.M.; Nonogaki, H. Longevity, Storage, and Deterioration. In *Seeds: Physiology of Development, Germination and Dormancy*, 3rd ed.; Bewley, J.D., Bradford, K.J., Hilhorst, H.W.M., Nonogaki, H., Eds.; Springer: New York, NY, USA, 2013; pp. 341–376, ISBN 978-1-4614-4693-4.
4. Bewley, J.D. Seed Germination and Dormancy. *Plant Cell* **1997**, *9*, 1055–1066. [CrossRef] [PubMed]
5. Ahmad, S.; Ahmad, R.; Ashraf, M.; Ashraf, M.; Waraich, E. Sunflower (*Helianthus annuus* L.) Response to Drought Stress at Germination and Seedling Growth Stages. *Pak. J. Bot.* **2009**, *41*, 647–654.
6. Anjum, S.A.; Ashraf, U.; Zohaib, A.; Tanveer, M.; Naeem, M.; Ali, I.; Tabassum, T.; Nazir, U. Growth and Developmental Responses of Crop Plants under Drought Stress: A Review. *Zemdirb.-Agric.* **2017**, *104*, 267–276. [CrossRef]
7. Chazen, O.; Hartung, W.; Neumann, P.M. The Different Effects of PEG 6000 and NaCI on Leaf Development Are Associated with Differential Inhibition of Root Water Transport. *Plant Cell Ent.* **1995**, *18*, 727–735. [CrossRef]
8. Hajihashemi, S.; Sofo, A. The Effect of Polyethylene Glycol-Induced Drought Stress on Photosynthesis, Carbohydrates and Cell Membrane in Stevia Rebaudiana Grown in Greenhouse. *Acta Physiol. Plant.* **2018**, *40*, 142. [CrossRef]
9. Kozlowski, T.T. *Germination Control. Metabolism, and Pathology*, 1st ed.; Elsevier: Amsterdam, The Netherlands, 1972; ISBN 978-0-12-424303-3.
10. Anderson, J.D.; Baker, J.E. Deterioration of Seeds during Aging. *Phytopathology* **1983**, *73*, 321–325. [CrossRef]
11. Rajjou, L.; Debeaujon, I. Seed Longevity: Survival and Maintenance of High Germination Ability of Dry Seeds. *Comptes Rendus Biol.* **2008**, *331*, 796–805. [CrossRef]
12. Abid, M.; Hakeem, A.; Shao, Y.; Liu, Y.; Zahoor, R.; Fan, Y.; Suyu, J.; Ata-Ul-Karim, S.T.; Tian, Z.; Jiang, D.; et al. Seed Osmopriming Invokes Stress Memory against Post-Germinative Drought Stress in Wheat (*Triticum aestivum* L.). *Environ. Exp. Bot.* **2018**, *145*, 12–20. [CrossRef]

13. Domokos-Szabolcsy, E.; Marton, L.; Sztrik, A.; Babka, B.; Prokisch, J.; Fari, M. Accumulation of Red Elemental Selenium Nanoparticles and Their Biological Effects in Nicotiniatabacum. *Plant Growth Regul.* **2012**, *68*, 525–531. [CrossRef]
14. Lerner, A.B.; Case, J.D.; Takahashi, Y.; Lee, T.H.; Mori, W. Isolation of Melatonin, the Pineal Gland Factor That Lightens Melanocytes. *J. Am. Chem. Soc.* **1958**, *80*, 2587. [CrossRef]
15. Dubbels, R.; Reiter, R.J.; Klenke, E.; Goebel, A.; Schnakenberg, E.; Ehlers, C.; Schiwara, H.W.; Schloot, W. Melatonin in Edible Plants Identified by Radioimmunoassay and by HPLC-MS. *J. Pineal Res.* **1995**, *18*, 28–31. [CrossRef]
16. Hattori, A.; Migitaka, H.; Iigo, M.; Itoh, M.; Yamamoto, K.; Ohtani-Kaneko, R.; Hara, M.; Suzuki, T.; Reiter, R.J. Identification of Melatonin in Plants and Its Effects on Plasma Melatonin Levels and Binding to Melatonin Receptors in Vertebrates. *Biochem. Mol. Biol. Int.* **1995**, *35*, 627–634. [PubMed]
17. Kolar, J.; Machackova, I.; Illnerova, H.; Prinsen, E.; van Dongen, W.; van Onckelen, H. Melatonin in Higher Plant Determined by Radioimmunoassay and Liquid Chromatography-Mass Spectrometry. *Biol. Rhythm. Res.* **1995**, *26*, 406–409.
18. Arnao, M.B. Phytomelatonin: Discovery, Content, and Role in Plants. *Adv. Bot.* **2014**, *2014*, e815769. [CrossRef]
19. Arnao, M.B.; Hernández-Ruiz, J. Phytomelatonin, Natural Melatonin from Plants as a Novel Dietary Supplement: Sources, Activities and World Market. *J. Funct. Foods* **2018**, *48*, 37–42. [CrossRef]
20. Arnao, M.B.; Cano, A.; Hernández-Ruiz, J. Phytomelatonin: An Unexpected Molecule with Amazing Performances in Plants. *J. Exp. Bot.* **2022**, *73*, 5779–5800. [CrossRef]
21. Arnao, M.B.; Hernández-Ruiz, J. Melatonin: A New Plant Hormone and/or a Plant Master Regulator? *Trends Plant Sci.* **2019**, *24*, 38–48. [CrossRef]
22. Arnao, M.B.; Hernández-Ruiz, J. Melatonin: Plant Growth Regulator and/or Biostimulator during Stress? *Trends Plant Sci.* **2014**, *19*, 789–797. [CrossRef]
23. Wei, J.; Li, D.-X.; Zhang, J.-R.; Shan, C.; Rengel, Z.; Song, Z.-B.; Chen, Q. Phytomelatonin Receptor PMTR1-Mediated Signaling Regulates Stomatal Closure in *Arabidopsis thaliana*. *J. Pineal Res.* **2018**, *65*, e12500. [CrossRef]
24. Byeon, Y.; Lee, H.Y.; Lee, K.; Park, S.; Back, K. Cellular Localization and Kinetics of the Rice Melatonin Biosynthetic Enzymes SNAT and ASMT. *J. Pineal Res.* **2014**, *56*, 107–114. [CrossRef] [PubMed]
25. Byeon, Y.; Yool Lee, H.; Choi, D.-W.; Back, K. Chloroplast-Encoded Serotonin N-Acetyltransferase in the Red Alga Pyropia Yezoensis: Gene Transition to the Nucleus from Chloroplasts. *J. Exp. Bot.* **2015**, *66*, 709–717. [CrossRef]
26. Tan, D.-X.; Reiter, R.J. An Evolutionary View of Melatonin Synthesis and Metabolism Related to Its Biological Functions in Plants. *J. Exp. Bot.* **2020**, *71*, 4677–4689. [CrossRef] [PubMed]
27. Zhao, D.; Yu, Y.; Shen, Y.; Liu, Q.; Zhao, Z.; Sharma, R.; Reiter, R.J. Melatonin Synthesis and Function: Evolutionary History in Animals and Plants. *Front. Endocrinol.* **2019**, *10*, 249. [CrossRef] [PubMed]
28. Hernández-Ruiz, J.; Giraldo-Acosta, M.; El Mihyaoui, A.; Cano, A.; Arnao, M.B. Melatonin as a Possible Natural Anti-Viral Compound in Plant Biocontrol. *Plants* **2023**, *12*, 781. [CrossRef] [PubMed]
29. Back, K. Melatonin Metabolism, Signaling and Possible Roles in Plants. *Plant J.* **2021**, *105*, 376–391. [CrossRef] [PubMed]
30. Back, K.; Tan, D.-X.; Reiter, R.J. Melatonin Biosynthesis in Plants: Multiple Pathways Catalyze Tryptophan to Melatonin in the Cytoplasm or Chloroplasts. *J. Pineal Res.* **2016**, *61*, 426–437. [CrossRef] [PubMed]
31. Tan, D.-X.; Hardeland, R.; Back, K.; Manchester, L.C.; Alatorre-Jimenez, M.A.; Reiter, R.J. On the Significance of an Alternate Pathway of Melatonin Synthesis via 5-Methoxytryptamine: Comparisons across Species. *J. Pineal Res.* **2016**, *61*, 27–40. [CrossRef] [PubMed]
32. Tan, D.-X.; Manchester, L.C.; Esteban-Zubero, E.; Zhou, Z.; Reiter, R.J. Melatonin as a Potent and Inducible Endogenous Antioxidant: Synthesis and Metabolism. *Molecules* **2015**, *20*, 18886–18906. [CrossRef] [PubMed]
33. Wang, S.-Y.; Shi, X.-C.; Wang, R.; Wang, H.-L.; Liu, F.; Laborda, P. Melatonin in Fruit Production and Postharvest Preservation: A Review. *Food Chem.* **2020**, *320*, 126642. [CrossRef] [PubMed]
34. Xu, T.; Chen, Y.; Kang, H. Melatonin is a potential target for improving post-harvest preservation of fruits and vegetables. *Front. Plant Sci.* **2019**, *10*, 1388. [CrossRef]
35. Arnao, M.B.; Hernández-Ruiz, J. Melatonin as a Regulatory Hub of Plant Hormone Levels and Action in Stress Situations. *Plant Biol.* **2021**, *23*, 7–19. [CrossRef]
36. Arnao, M.B.; Hernández-Ruiz, J. Melatonin and Its Relationship to Plant Hormones. *Ann. Bot.* **2018**, *121*, 195–207. [CrossRef] [PubMed]
37. Gu, Q.; Chen, Z.; Yu, X.; Cui, W.; Pan, J.; Zhao, G.; Xu, S.; Wang, R.; Shen, W. Melatonin Confers Plant Tolerance against Cadmium Stress via the Decrease of Cadmium Accumulation and Reestablishment of microRNA-Mediated Redox Homeostasis. *Plant Sci.* **2017**, *261*, 28–37. [CrossRef] [PubMed]
38. Reiter, R.J.; Rosales-Corral, S.; Tan, D.X.; Jou, M.J.; Galano, A.; Xu, B. Melatonin as a Mitochondria-Targeted Antioxidant: One of Evolution's Best Ideas. *Cell. Mol. Life Sci.* **2017**, *74*, 3863–3881. [CrossRef] [PubMed]
39. Moustafa-Farag, M.; Almoneafy, A.; Mahmoud, A.; Elkelish, A.; Arnao, M.B.; Li, L.; Ai, S. Melatonin and Its Protective Role against Biotic Stress Impacts on Plants. *Biomolecules* **2020**, *10*, 54. [CrossRef] [PubMed]
40. Posmyk, M.M.; Balabusta, M.; Wieczorek, M.; Sliwinska, E.; Janas, K.M. Melatonin Applied to Cucumber (*Cucumis sativus* L.) Seeds Improves Germination during Chilling Stress. *J. Pineal Res.* **2009**, *46*, 214–223. [CrossRef] [PubMed]
41. Posmyk, M.M.; Kuran, H.; Marciniak, K.; Janas, K.M. Presowing Seed Treatment with Melatonin Protects Red Cabbage Seedlings against Toxic Copper Ion Concentrations. *J. Pineal Res.* **2008**, *45*, 24–31. [CrossRef] [PubMed]

42. Bewley, J.D.; Bradford, K.J.; Hilhorst, H.W.M.; Nonogaki, H. *Seeds: Physiology of Development, Germination and Dormancy*, 3rd ed.; Springer: New York, NY, USA, 2013; ISBN 978-1-4614-4692-7.
43. Jincya, M.; Prasad, V.B.R.; Jeyakumara, P.; Senthila, A.; Manivannan, N. Evaluation of Green Gram Genotypes for Drought Tolerance by PEG (Polyethylene Glycol) Induced Drought Stress at Seedling Stage. *Legume Res.* **2019**, *44*, 684–691. [CrossRef]
44. Megala, R.; Kalarani, M.K.; Jeyakumar, P.; Senthil, N.; Pushpam, R.; Umapathi, M. Standardization of Optimum Melatonin Concentration for Drought Tolerance at Germination and Early Development Stage in Rice (CO-54). *J. Appl. Nat. Sci.* **2022**, *14*, 1022–1030. [CrossRef]
45. Wang, G.; Yang, Y.; Kong, Y.; Ma, R.; Yuan, J.; Li, G. Key Factors Affecting Seed Germination in Phytotoxicity Tests during Sheep Manure Composting with Carbon Additives. *J. Hazard. Mater.* **2022**, *421*, 126809. [CrossRef]
46. Hernández-Herrera, R.M.; Santacruz-Ruvalcaba, F.; Ruiz-López, M.A.; Norrie, J.; Hernández-Carmona, G. Effect of Liquid Seaweed Extracts on Growth of Tomato Seedlings (*Solanum lycopersicum* L.). *J. Appl. Phycol.* **2014**, *26*, 619–628. [CrossRef]
47. Kamran, M.; Wang, D.; Xie, K.; Lu, Y.; Shi, C.; EL Sabagh, A.; Gu, W.; Xu, P. Pre-Sowing Seed Treatment with Kinetin and Calcium Mitigates Salt Induced Inhibition of Seed Germination and Seedling Growth of Choysum (*Brassica rapa* Var. *Parachinensis*). *Ecotoxicol. Environ. Saf.* **2021**, *227*, 112921. [CrossRef] [PubMed]
48. Amin, H.; Arain, B.A.; Amin, F.; Surhio, M.A. Phytotoxicity of Chromium on Germination, Growth and Biochemical Attributes of *Hibiscus esculentus* L. *Am. J. Plant Sci.* **2013**, *4*, 2431–2439. [CrossRef]
49. García-Sánchez, S.; Cano, A.; Hernández-Ruiz, J.; Arnao, M.B. Effects of Temperature and Light on the Germination-Promoting Activity by Melatonin in Almond Seeds without Stratification. *Agronomy* **2022**, *12*, 2070. [CrossRef]
50. Hernández-Ruiz, J.; Cano, A.; Arnao, M.B. Melatonin: A Growth-Stimulating Compound Present in Lupin Tissues. *Planta* **2004**, *220*, 140–144. [CrossRef] [PubMed]
51. Korkmaz, A. The Effects of Seed Treatment with Melatonin on Germination and Emergence Performance of Pepper Seeds under Chilling Stress. *Tarim Bilim. Derg.-J. Agric. Sci.* **2017**, *23*, 167–176.
52. Lv, Y.; Pan, J.; Wang, H.; Reiter, R.J.; Li, X.; Mou, Z.; Zhang, J.; Yao, Z.; Zhao, D.; Yu, D. Melatonin Inhibits Seed Germination by Crosstalk with Abscisic Acid, Gibberellin, and Auxin in Arabidopsis. *J. Pineal Res.* **2021**, *70*, e12736. [CrossRef] [PubMed]
53. Simlat, M.; Ptak, A.; Skrzypek, E.; Warchoł, M.; Morańska, E.; Piórkowska, E. Melatonin Significantly Influences Seed Germination and Seedling Growth of Stevia Rebaudiana Bertoni. *PeerJ* **2018**, *6*, e5009. [CrossRef] [PubMed]
54. Xiao, S.; Liu, L.; Wang, H.; Li, D.; Bai, Z.; Zhang, Y.; Sun, H.; Zhang, K.; Li, C. Exogenous Melatonin Accelerates Seed Germination in Cotton (*Gossypium hirsutum* L.). *PLoS ONE* **2019**, *14*, e0216575. [CrossRef] [PubMed]
55. Zhang, M.; Liu, S.; Tong, S.; Zhang, D.; Qi, Q.; Wang, Y.; Wang, X.; An, Y.; Lu, X. Effects of Melatonin Priming on Suaeda Corniculata Seed Germination, Antioxidant Defense, and Reserve Mobilization: Implications for Salinized Wetland Restoration. *Front. Ecol. Evol.* **2022**, *10*, 941032. [CrossRef]
56. Hernández-Ruiz, J.; Cano, A.; Arnao, M.B. Melatonin Acts as a Growth-Stimulating Compound in Some Monocot Species: Melatonin as Growth Promoter in Plants. *J. Pineal Res.* **2005**, *39*, 137–142. [CrossRef]
57. Chen, Q.; Qi, W.; Reiter, R.J.; Wei, W.; Wang, B. Exogenously Applied Melatonin Stimulates Root Growth and Raises Endogenous Indoleacetic Acid in Roots of Etiolated Seedlings of *Brassica juncea*. *J. Plant Physiol.* **2009**, *166*, 324–328. [CrossRef]
58. Li, L.; Verstraeten, I.; Roosjen, M.; Takahashi, K.; Rodriguez, L.; Merrin, J.; Chen, J.; Shabala, L.; Smet, W.; Ren, H.; et al. Cell Surface and Intracellular Auxin Signalling for H+ Fluxes in Root Growth. *Nature* **2021**, *599*, 273–277. [CrossRef] [PubMed]
59. Arnao, M.B.; Hernández-Ruiz, J. Melatonin Promotes Adventitious- and Lateral Root Regeneration in Etiolated Hypocotyls of *Lupinus albus* L. *J. Pineal Res.* **2007**, *42*, 147–152. [CrossRef] [PubMed]
60. Ren, S.; Rutto, L.; Katuuramu, D. Melatonin Acts Synergistically with Auxin to Promote Lateral Root Development through Fine Tuning Auxin Transport in *Arabidopsis thaliana*. *PLoS ONE* **2019**, *14*, e0221687. [CrossRef] [PubMed]
61. Zhang, M.; Gao, C.; Xu, L.; Niu, H.; Liu, Q.; Huang, Y.; Lv, G.; Yang, H.; Li, M. Melatonin and Indole-3-Acetic Acid Synergistically Regulate Plant Growth and Stress Resistance. *Cells* **2022**, *11*, 3250. [CrossRef] [PubMed]
62. Merritt, D.J.; Senaratna, T.; Touchell, D.H.; Dixon, K.W.; Sivasithamparam, K. Seed Ageing of Four Western Australian Species in Relation to Storage Environment and Seed Antioxidant Activity. *Seed Sci. Res.* **2003**, *13*, 155–165. [CrossRef]
63. Rajjou, L.; Lovigny, Y.; Groot, S.P.C.; Belghazi, M.; Job, C.; Job, D. Proteome-Wide Characterization of Seed Aging in Arabidopsis: A Comparison between Artificial and Natural Aging Protocols. *Plant Physiol.* **2008**, *148*, 620–641. [CrossRef] [PubMed]
64. Ratajczak, E.; Małecka, A.; Bagniewska-Zadworna, A.; Kalemba, E.M. The Production, Localization and Spreading of Reactive Oxygen Species Contributes to the Low Vitality of Long-Term Stored Common Beech (*Fagus sylvatica* L.) Seeds. *J. Plant Physiol.* **2015**, *174*, 147–156. [CrossRef] [PubMed]
65. Wang, L.; Tanveer, M.; Wang, H.; Arnao, M.B. Melatonin as a Key Regulator in Seed Germination under Abiotic Stress. *J. Pineal Res.* **2024**, *76*, e12937. [CrossRef] [PubMed]
66. Korkmaz, A.; Köklü, S.; Yakupoglu, G. Investigating the Effects of Melatonin Application on the Ageing Process of Pepper Seeds. In Proceedings of the Acta Horticulturae; International Society for Horticultural Science (ISHS), Leuven, Belgium, 30 June 2018; pp. 9–16.
67. Yakupoglu, G.; Köklü, S.; Karaca, A.; Düver, E.; Reiter, J.R.; Korkmaz, A. Fluctuations in Melatonin Content and Its Effects on the Ageing Process of Lettuce Seeds during Storage. *Acta Sci. Pol. Hortorum Cultus* **2021**, *20*, 77–88. [CrossRef]

68. Zhang, H.; Zhang, N.; Yang, R.; Wang, L.; Sun, Q.; Li, D.; Cao, Y.; Weeda, S.; Zhao, B.; Ren, S.; et al. Melatonin Promotes Seed Germination under High Salinity by Regulating Antioxidant Systems, ABA and GA $_4$ Interaction in Cucumber (*Cucumis sativus* L.). *J. Pineal Res.* **2014**, *57*, 269–279. [CrossRef]
69. Rosińska, A.; Andrzejak, R.; Kakkerla, V. Effect of Osmopriming with Melatonin on Germination, Vigor and Health of Daucus Carota L. Seeds. *Agriculture* **2023**, *13*, 749. [CrossRef]
70. Zhang, M.; He, S.; Qin, B.; Jin, X.; Wang, M.; Ren, C.; Cao, L.; Zhang, Y. Exogenous Melatonin Reduces the Inhibitory Effect of Osmotic Stress on Antioxidant Properties and Cell Ultrastructure at Germination Stage of Soybean. *PLoS ONE* **2020**, *15*, e0243537. [CrossRef] [PubMed]
71. Guo, Y.; Li, D.; Liu, L.; Sun, H.; Zhu, L.; Zhang, K.; Zhao, H.; Zhang, Y.; Li, A.; Bai, Z.; et al. Seed Priming with Melatonin Promotes Seed Germination and Seedling Growth of *Triticale hexaploide* L. Under PEG-6000 Induced Drought Stress. *Front. Plant Sci.* **2022**, *13*, 932912. [CrossRef] [PubMed]

seeds

Review

Improvement of Seed Quality by Priming: Concept and Biological Basis

Françoise Corbineau [1,*], Nesrin Taskiran-Özbingöl [2] and Hayat El-Maarouf-Bouteau [3,4]

1. Biologie des Semences, UMR7622 Biologie du Développement, IBPS, Sorbonne-Université, 4 place Jussieu, CEDEX 05, 75252 Paris, France
2. Lycée des Métiers de l'Horticulture et du Paysage Jeanne Baret, 93100 Montreuil, France
3. UFR 927, Sorbonne-Université, 4 place Jussieu, CEDEX 05, 75252 Paris, France
4. Institut Jean-Pierre Bourjin, INRAE, CNRS, AgroParisTech, Université Paris-Saclay, Route de Saint-Cyr, 78000 Versailles, France

* Correspondence: francoise.corbineau@sorbonne-universite.fr

Abstract: Presoaking seeds in water (hydropriming) or in a solution, usually of polyethylene glycol (PEG) or various salts at low water potential (osmopriming), has been demonstrated to improve the germination of seeds of numerous species including vegetables (carrot, celery, leek, lettuce, tomato), floral plants (cyclamen, primrose, pansy) and others (sugar beet, rape, soybean, sunflower). This treatment allows the germination stricto sensu to occur but prevents the radicle protrusion. Germination of primed seeds is more rapid and uniform than that of unprimed ones. Primed seeds germinate in a wider range of temperatures and are less sensitive to oxygen deprivation. Interestingly, priming also improves the germination of aged seeds. The stimulatory effect of priming persists after redrying and often during storage; however, primed seeds often deteriorate faster during storage or accelerated aging than unprimed ones. A better understanding of the mechanisms involved during priming allows us to suggest markers of the effectiveness of priming. Among these markers, ethylene production during imbibition, cell-cycle processes (DNA replication, ß-tubulin), soluble sugar metabolism (raffinose family oligosaccharides, in particular), reactive oxygen species scavenging through antioxidant systems and energy metabolism are correlated to seed vigor. Global approaches (proteomic, metabolomic or transcriptomic) could also result in the identification of new markers.

Keywords: seed quality; hydropriming; osmopriming; markers of priming; energy metabolism; antioxidant defense system; cell cycle; omics

Citation: Corbineau, F.; Taskiran-Özbingöl, N.; El-Maarouf-Bouteau, H. Improvement of Seed Quality by Priming: Concept and Biological Basis. *Seeds* **2023**, *2*, 101–115. https://doi.org/10.3390/seeds2010008

Academic Editors: José Antonio Hernández Cortés, Gregorio Barba-Espín and Pedro Diaz-Vivancos

Received: 27 January 2023
Revised: 13 February 2023
Accepted: 15 February 2023
Published: 20 February 2023

1. Introduction

Successful stand establishment requires high-quality seeds, i.e., seeds that (1) all germinate, (2) germinate quickly and simultaneously, (3) give rise to normal and vigorous seedlings, (4) display low sensitivity to external factors (temperature, oxygen availability, water potential of the soil) and lastly, (5) germinate in a wide range of environmental conditions [1–5]. It is well known that acquisition of seed quality occurs during seed development and the maturation phase [6], and that seed quality can be improved by breeding and selection, two fundamental approaches. For example, quantitative trait loci (QTLs) related to germination rate have been detected in sunflower, rape and *Medicago truncatula* seeds [7–10]. Seed companies may also enhance seed quality at different steps of the seed production, by improving the methods of harvest but often by post-harvest treatments such as cleaning, sorting, coating, priming and controlling the storage conditions [1,11–13]. The treatments can be grouped in 3 groups: (1) conditioning (cleaning, purification, fractionation using size and density grading, color sorting, polishing and scarification), (2) seed protection by applying active compounds (fungicides and/or insecticides) and (3) seed invigoration, also called physiological enhancement, such as priming. Three mains strategies are used for improving seed quality by the priming technology [1,3,4,11–13]: seed hydration with water for various durations (hydropriming); submersion in solutions of osmotica (osmopriming); mixing seeds with moist solid particle materials (matrix priming).

This review will be focused mainly on hydropriming and osmopriming, but other techniques are in development such as, for example, biopriming (hydration with microorganisms) and nanopriming (hydration with agents such as nanoparticles of silver and zinc oxide) [4,5,14,15].

The objective of this review is to indicate the main priming technologies, to describe some beneficial effects of the treatment in relation to temperature and oxygen supply and to specify the influence of the priming conditions on the efficiency of the treatment. It is also to better understand the cellular, biochemical and molecular mechanisms associated with these treatments, which could suggest markers of the efficiency of priming.

2. Main Conventional Seed Priming Techniques

Seed priming, a technique now used commercially, has been demonstrated to improve the germination of seeds of numerous vegetable plants (leek, tomato, pepper, onion and carrot) [1,16–21] and the production of potted or bedding ornamental plants such as cyclamen, begonia, pansy and primrose, as well as for large volume of field crops such as sugar beet and turf grasses [1,3,4,22].

Hydropriming is a simple technique in which seeds are immersed in water for a specific period that does not allow radicle protrusion and permits seeds to dry back to their initial water content. However, it is difficult to avoid the radicle growth since hydropriming is a non-controlled water uptake [23]. Drum priming is a key technique that allows a controlled increase in water uptake and seed imbibition during the treatment by regularly measuring seed mass and the volume of water required to control seed hydration [24,25].

Osmopriming corresponds to seed submersion with aerated solutions of low water potential (usually −1.0 to −2.0 MPa): polyethylene-glycol (PEG), mannitol and sorbitol or different salts ($NaNO_3$, $MgCl_2$, KH_2PO_4, $KH(PO_4)_2$, K_3PO_4, KCl, KNO_3, $Ca(NO_3)_2$) [1,3,4,26,27]. The duration of priming depends on the species and varies from 1–2 days (lettuce, rice, sorghum) up to 5–7 days (tomato, sugar beet), 7–10 days (cauliflower, carrot, fennel, primrose, pansy) or 14 days (celery, leek). The seed industry generally dries back primed seeds before storage and sowing.

Figure 1 illustrates the priming process as related to water content and some metabolic changes [5,15,27,28]. Priming (hydro- or osmopriming) consists of partial hydration of the seed population, allowing the phase II of germination sensu stricto to occur but preventing the radicle emergence (phase III, growth) associated with the loss of desiccation tolerance. The seed moisture content (MC) is maintained during the priming treatment at 40–45% of the fresh weight basis which corresponds to an MC at about 90–95% that allows radicle emergence [1,3,4,26,27]. Treated seeds are redried to their initial moisture content and stored before sowing. During the imbibition phase, controlled water uptake allows the protein synthesis and induces respiration activity. Phase II, called "pregermination" or "activation", is associated with several metabolic processes including protein synthesis, respiration, metabolism of sugar, etc. Seeds are then dehydrated in order to postpone seed sowing.

Nanopriming and biopriming are advanced methods that have shown promising beneficial effects in agriculture. The nanoagents are silver and zinc oxide nanoparticles [29], and biopriming integrates biological aspects by inoculating seeds with beneficial microorganisms controlling seed-borne pathogens and biological treatment corresponding to partial seed imbibition [15,30].

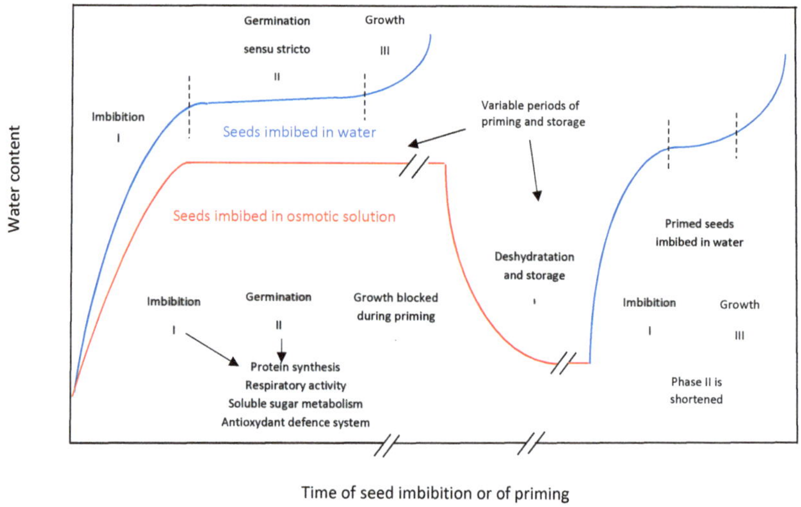

Figure 1. Seed priming process as related to seed water content. Priming consists of partial seed imbibition to a point where germination (phase II, germination sensu stricto) occurs but is not completed by radicle growth. The moisture content of the seeds is maintained during priming at 40–45% fresh weight basis, corresponding to about 90–95% of the moisture content that allow radicle growth. During phase I (imbibition), controlled water uptake allows protein synthesis and induces respiratory activity. Phase II is associated to various physiological, biochemical and molecular activities such as protein synthesis, respiratory activity, metabolism of soluble sugars and repair processes, but the radicle emergence is prevented. In seed company, seeds are dehydrated after priming and stored in order to postpone seed sowing. After sowing, primed seed germination rate is improved and phase II is shortened. "//" indicates a change in the x-axis scale depending on the duration of priming or the duration of storage of primed seeds.

3. Beneficial Effects of Priming

3.1. Seed Sensitivity to Temperature and Oxygen

Priming strongly improves the subsequent germination of seeds in water in a wide range of temperatures (Figure 2). For example, without priming, leek seeds germinate only in a narrow range of temperatures (10–20 °C), the thermal optimum being 15–20 °C, while they germinate easily between 5 and 40 °C after osmopriming (Figure 2) [21,31]. Osmoprimed tomato seeds germinate in higher percentages at 35 °C than unprimed ones and are able to germinate at 10 °C, whereas control seeds do not germinate at low temperatures (Figure 2) [32–34]. After osmopriming, the germination of carrot seeds is improved at low temperatures (5–10 °C) (Figure 2) [17,18,31], i.e., when sowing occurs early in spring. In case of a mixture of seeds from various genotypes, Figure 3 shows that priming could homogenize the germination of *Primula* seeds from genotypes with blue, carmine and yellow flowers, which is a good tool to produce seedlings in greenhouse conditions and reduced time due to the manual planting out of seedlings in order to obtain homogeneous and simultaneous development.

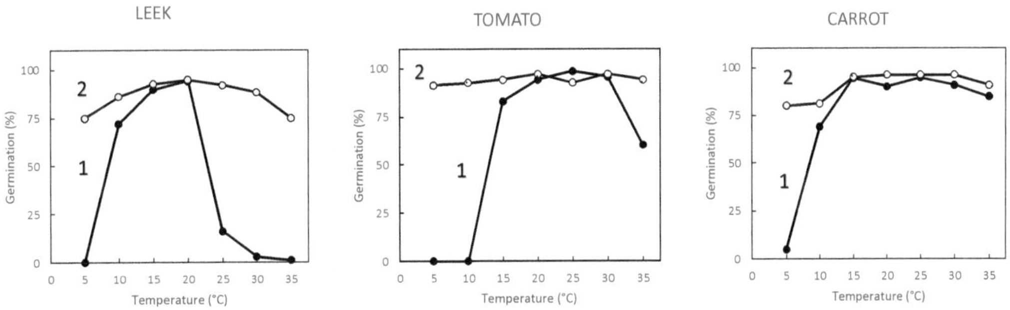

Figure 2. Effects of temperature on the germination percentages obtained after 7 days with unprimed seeds (1) and primed seeds (2) of leek, tomato and carrot. 1, control non-primed seeds; 2, seeds pretreated for 14 days (leek), 9 days (carrot) and 7 days (tomato) at 15 °C in the presence of a solution of polyethylene glycol at −1.0 MPa (tomato) or −1.5 MPa (leek, carrot). Means of 4 replicates of 50 seeds. Modified from [2,21,32,34].

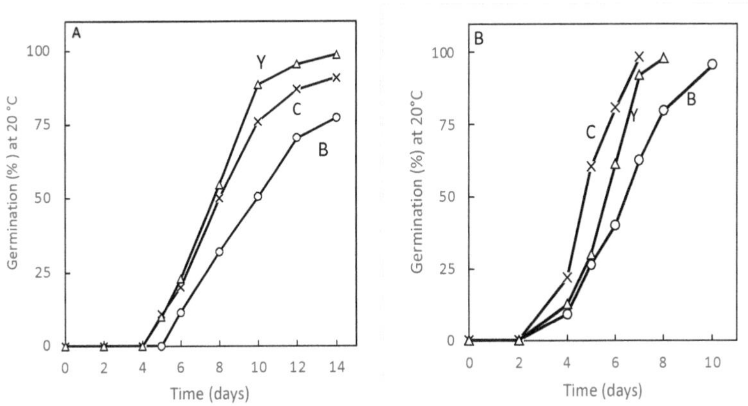

Figure 3. Germination at 20 °C of *Primula* seeds from genotype with blue (B), carmine (C) and yellow (Y) flowers. (**A**), seeds non-primed; (**B**), seeds primed on PEG solutions. Means of 4 replicates of 50 seeds. From Corbineau (unpublished data).

Primed seeds are also less sensitive to oxygen deprivation than control unprimed ones [2,21,32,34–37]. Table 1 shows that primed seeds of carrot, Lamb's lettuce, leek, sunflower and tomato germinate faster and in higher percentages in low oxygen concentrations (5–15%) than non-primed seeds.

Table 1. Effect of priming on seed sensitivity to oxygen tension. Seeds are not primed or are osmoprimed for 7 days at 15 °C in the presence of PEG-8000 solution at −1 MPa (tomato, lamb's lettuce), 5 days at 15 °C with PEG-8000 solution at −2 MPa (sunflower), 9 days (carrot) and 14 days (leek) at 15 °C in the presence of PEG-8000 solution at −1.5 MPa. Germination percentages are counted after 4 days at 20 °C (carrot, leek, lamb's lettuce), after 2 days (tomato) or 4 days (sunflower) at 25 °C. Means of four replicates. Modified from [19,21,32,34–37].

Species	Seed Treatment	Germination (%) in Atmosphere Containing 1 to 21% Oxygen					
		1%	3%	5%	10%	15%	21%
Carrot [37]	Non-primed	0	0	0	13.1	23.0	70.0
	Primed	0	5.5	34.6	78.5	84.6	94.1
Lamb's lettuce [19,34]	Non-primed	0	0	4.2	5.3	10.5	51.1
	Primed	0	4.4	23.0	92.2	97.5	98.1
Leek [21,34]	Non-primed	0	0	3.1	20.2	52.2	53.2
	Primed	0	11.4	34.4	85.2	93.3	95.2
Sunflower [35,36]	Non-primed	4.6	40.7	55.6	79.6	92.5	100
	Primed	10.2	75.6	95.3	100	100	100
Tomato [32,34,37]	Non-primed	0	0	0	0	20.7	48.4
	Primed	2.2	10.1	50.1	76.7	92.4	95.5

Experiments in field conditions have also shown that priming increases the seedling emergence in suboptimal conditions of sowing and mean plant weight [18]. The time to obtain 50% emergence was reduced by 2–4 days in Brussels sprouts and cabbage seeds [38]. Seed priming not only enhances the germination in a large range of conditions but also enhances stress tolerance of the plants [1,3,5,15].

3.2. Germination of Aged Seeds

Priming improves the germination of low vigor or aged seeds of various species such as wheat [39], cauliflower [40], tomato [41], sunflower [42], pepper [43] and rape [34,44]. For example, after aging at 45 °C and 100% relative humidity for 31 h, rape seeds are still viable, but their germination rate is reduced compared to the control unaged seeds [34]; however, osmopriming at 25 °C in a PEG solution at −2 MPa enhances the germination of aged seeds. This improving effect increases with the duration of the treatment, and, after 6 days of priming, aged seeds germinate as well as the unaged ones. The more the seeds are aged the longer they must be primed for restoring their initial germination ability [34,44]. In the case of sunflower seeds, the re-invigoration during priming of seeds aged at 45 °C for 5 days is associated with a decrease in lipid peroxidation and the recovery of the detoxifying enzyme (superoxide dismutase, catalase, glutathione reductase) activities [42].

3.3. Examples of Priming Beneficial Effects on Several Species

The stimulatory effects of priming depend on the conditions (particularly temperature, water potential and oxygen availability) and the duration of the treatment [3,5,11,12,15]. Water potential during osmopriming varies between −0.5 and −2 MPa depending on the species, but generally the moisture content of the seeds is maintained at around 40–45% fresh weight basis, which is lower than the moisture content that would allow radicle protrusion [27]. The temperature range and oxygen concentrations which are effective during priming are similar to those which allow the germination of unprimed seeds, which demonstrates that priming corresponds to the realization of the germination stricto sensu (phase II of the germination process). For example, the optimal temperature for priming and germination of unprimed seeds is around 25 °C for tomato [32,33], 20–25 °C for sunflower [34] and 10–15 °C for leek [34]. Table 2 gives various examples of the priming treatment procedures on different species and their effects on germination and sensitivity to temperature or hypoxia.

Table 2. Optimal priming (hydropriming and osmopriming) procedures that improve the germination of vegetable, horticultural and crops species (compiled from [16–79]).

Species	Optimum Priming Treatment	Beneficial Effects
	Horticultural species	
Primula acaulis (primrose)	8–10 days at 20 °C with PEG-8000 at −1.5 MPa (Corbineau, unpublished)	- Improvement of the germination rate - Germination in a larger temperature range
Primula obconica	8–10 days at 20 °C with PEG-8000 at −1.5 MPa (Corbineau, unpublished)	- Improvement of the germination rate - Germination in a larger temperature range
Viola x wittrockiana (pansy)	PEG-8000 at −2.5 MPa at 25 °C (Corbineau, unpublished)	- Enhancement of the germination at 5 °C
	Vegetable species	
Allium porrum (leek)	14 days at 15 °C with PEG solution at −1 MPa [21,27,34]	- Enhancement of the germination rate - Improvement of germination at temperature higher than 20 °C
Apium graveolens (celery)	10–14 days with PEG at 15 °C at −1.2 MPa [17,18,45]	- Growth of the embryo associated with an increase in the germination rate
Brassica oleracea (cauliflower)	Hydropriming: incubation for 2–4 days with water content about 40% fresh matter Osmopriming: 7 days with PEG-8000 at 20 °C at −1.5 or −2 MPa [40,46].	- Enhancement of the germination rate - Improvement of the germination at 5 °C
Daucus carota (carrot)	Osmopriming: 3–7 days at 20 °C with PEG-8000 solution at −1.0 to 1.5 MPa [37,47]	- Enhancement of the germination at low temperatures
Capsicum annuum (pepper)	12 days at 20 °C with PEG-8000 solution at −1.1 to −1.5 MPa [45,48,49]	- Enhancement of the germination rate - Improvement of germination in a wide range of temperature
Foeniculum vulgare (fennel)	5–7 days at 20 °C with PEG-8000 solution at −1.5 MPa (Özbingöl, unpublished)	- Enhancement of the germination at low temperatures (5–10 °C)
Lactica sativa (lettuce)	2 days at 15 °C with PEG-8000 at −1.2 or −1.3 MPa [16,45,50–54]	- Enhancement of germination rate - Improvement of germination at temperature higher than 25 °C
Lycopersicon esculentum (tomato)	5–7 days at 15–25 °C with PEG-8000 solution at −1 MPa to −1.5 MPa, or in a KNO_3 solution at −1.4 MPa [32,33,45,55,56],	- Enhancement of the germination rate - Improvement of germination at low temperature - Reduction of the lag time - Increase in water uptake - Reduction of seed sensitivity to hypoxia
Spinacia oleracea (spinach)	8 days at 15 °C with PEG at −0.6 MPa [57,58]	- Enhancement of seed germination
Valerianella olitoria (lamb's lettuce)	Hydro priming: 40 h at 20 °C [19]	- Reduction of seed sensitivity to hypoxia

Table 2. *Cont.*

Species	Optimum Priming Treatment	Beneficial Effects
	Crop species	
Beta vulgaris (sugar beet)	Hydropriming: 2 to 5 days at 20–25 °C Osmopriming 2 to 7 days at 25 °C in PEG-8000 solution at −2 MPa [22,59–61]	- Improvement of the germination at low temperature (5–10 °C)
Brassica napus (rape)	PEG at −1.2 MPa at 20 °C [62,63]	- Improvement of salinity tolerance
Glycine max (soybean)	1–2 weeks at 20 °C with PEG-8000 at −1.5 MPa [64–66]	- Enhancement the tolerance to chilling - Reduction of chilling sensitivity
Helianthus annuus (sunflower)	Hydropriming: 18 h at 25 °C [67] Osmopriming: 3 to 7 days at 15 °C with PEG-8000 solution at −1.5–2.0 MPa [35,36,68,69]	- Improvement of germination at temperature lower than 10–15 °C - Increase in respiration - Reduction of seed sensitivity to oxygen deprivation - Enhancement of the ACC conversion to ethylene - Stimulation in catalase and gluthation reductase during priming
Hordeum vulgare (barley)	Hydropriming: 30 °C with 40–52% moisture content [70]	- Induction of the cell cycle - Decrease in ABA content in the embryo
Oryza sativa (rice)	Hydropriming: 12 h in water [71] Osmopriming 12–24 h in the presence of 50–75 mM NaCl, Salicylic acid or polyamines [72–74]	- Enhancement of the oxidative and anti-oxidative mechanisms - Improvement of the tolerance to salinity and drought
Sorghum bicolor (sorghum)	Osmopriming: 48 h with PEG solution at 18 °C [75]	- Enhancement of antioxidant activities - Increase tolerance of plants to drought conditions
Triticum aestivum (wheat)	Hydropriming: 24 h in water Osmopriming: with CaCl2 or KCl solutions at −1.25 MPa [76–78]	- Improvement of chilling tolerance
	Model plant	
Arabidopsis thaliana (arabidopsis)	Hydropriming: 1 day at 25 °C Osmopriming: 5 to 7 days at 20 °C in a PEG-8000 solution at 0.75 MPa [79]	- Enhancement of the germination rate

To be efficient, the priming treatment requires more than 5% oxygen in the atmosphere (Figure 4) [33,34], indicating that metabolic processes are necessary for the syntheses associated with priming [27].

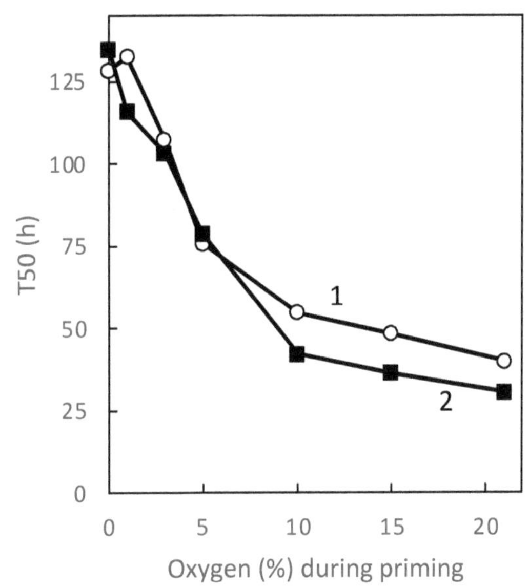

Figure 4. Effects of oxygen tension during priming for 7 days at 15 °C (1) or 25 °C (2) on a PEG solution at −1.0 MPa on the time to obtain 50% germination (T_{50}) with tomato seeds (cv Elko) transferred on water at 15 °C. Mean of 3 replicates of 50 seeds. T50 of dry non-primed seeds was 129 h. From [33,34].

4. Markers of Priming

A better understanding of the metabolic, biochemical and molecular mechanisms involved in the enhancement of germination after priming would allow us to suggest various biochemical and molecular markers of priming treatment. Among these, respiration and ethylene production during seed imbibition [80,81], protein, RNA and DNA synthesis [27,28], DNA replication [33,39,48,82] or β-tubulin accumulation [55,83] involved in the cell cycle regulation, soluble sugar metabolism and activity of the antioxidant defense systems [58,68,72,80] are promising for evaluating the efficiency of priming treatments.

4.1. Respiration and Ethylene Synthesis

The respiratory activity increases during seed priming, this effect being associated with seed imbibition. Various studies [32,35,36,81,84–86] have shown that the respiratory activity is stimulated during priming and after transfer of primed seeds onto water. Priming also stimulates ATP synthesis and the ATP/ADP ratio during the imbibition phase. This stimulatory effect rises with increasing duration of osmotic treatment. In tomato seeds [81], the beneficial effect of priming increases with increasing energy metabolism; it is optimal when the energy charge (EC) and ATP/ADP ratio are higher than about 0.75 and 1.7, respectively.

The ability of the seeds to convert 1-aminocyclopropane-1-carboxylic acid (ACC), the direct precursor of ethylene, is a good indicator of membrane properties since it is mediated by ACC oxidase, the in vivo activity of which depends on membrane integrity. It is a good marker of seed vigor in various species such as lettuce, cabbage, tomato, sweet-corn [87] and sunflower [36]. ACC-dependent ethylene production is also well correlated with the efficiency of priming treatment in sunflower [36] and carrot [2,80].

4.2. *Soluble Sugars and Oxidative Status*

In terms of seed vigor, soluble sugar metabolism and the capacity to scavenge reactive oxygen species (ROS) seem to be of a particular interest.

Priming is often associated with changes in soluble sugar content. There is a sublinear relationship between sucrose and oligosaccharide contents and the sucrose/oligosaccharide ratio, and the efficiency of priming evaluated by the T_{50} at 10 °C in carrot [2,80] and the germination % obtained after 4 days at 10 °C in fennel (Özbingöl, unpublished data). In tomato seeds, the ratio raffinose/sucrose is around 0.48 in dry unprimed seeds and decreases down 0.20–0.22 or 0.11–0.13 after 3 days of priming at 15 and 25 °C, respectively, and down 0.10–0.06 after 7 days of priming [88]. An increase in sucrose was also observed in impatiens and cucumber germinated seeds placed on a PEG solution at −1.5 MPa [89].

Reactive oxygen species play a critical role in sensing the environmental conditions and are key regulators of the germination process [90,91]. In rice, different priming techniques increase the metabolites/non-enzymatic antioxidant contents (total sugars, total phenolics, free amino acids, proline, ascorbate and glutathione) as well as activities of antioxidant enzymes (superoxide dismutase, catalase, ascorbate peroxidase and guaiacol peroxidase), thus reducing oxidative stress damage. The catalase activity is also sublinearly correlated with the germination rate of sunflower [42,68,69]. Osmopriming strongly enhances SOD and CAT activity and thus improves the antioxidant defense of the cells [68]. In addition, osmopriming of aged seeds completely restores the initial rate of germination [69].

4.3. *Cell Cycle Regulation*

Studies using flow cytometry have demonstrated that DNA replication is initiated in the radicle tip cells during osmopriming of tomato seeds [33,48,82,92,93]. An increase in 4C DNA content is also observed in sugar beet seeds [94]. The amounts of 4C nuclei reach 28.8% after 7 days of priming at 25 °C and on a PEG solution at −1.0 MPa in pepper and in several tomato cultivars, such as Elko, Agata and San Marzano [33,48,82]. The DNA synthesis is inversely related to water potential of the osmoticum [33] and Figure 5 shows that a positive linear relationship exists between temperature of priming up to 25 °C and the 4C signals or the 4C/2C ratio, and a negative linear relationship exists at higher temperatures. To promote the germination, the priming treatment requires at least 5% oxygen; similar oxygen concentration is also required for DNA replication [33].

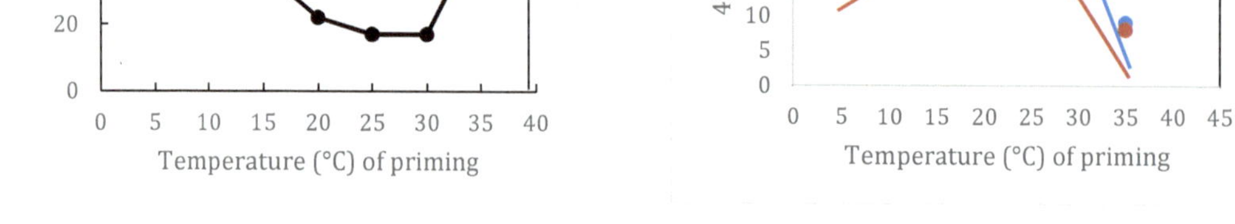

Figure 5. Effect of temperature of priming (7 days with a PEG solution at −1 MPa) on the germination rate of tomato seeds at 15 °C (expressed as the time to obtain 50% germination: T50) (**A**) and on the percentage of 4C nuclei (red) and the 4C/2C ratio (blue) (**B**). From [33,34].

4.4. Global Analyses Using Omics

Omics analyses have been extensively performed for better understanding the global process of priming. However, it is difficult to distinguish between newly synthetized priming-induced genes or proteins from those corresponding to stored mRNA or those resulting from turnover changes. mRNA and ribosome synthesis as well as translation initiation or transcription factors candidates were up-regulated during osmopriming in transcriptome in rice or *Brassica oleracea* seeds or during hydropriming in proteome of durum wheat, suggesting the activation of de novo transcription and translation [95–97] In response to PEG treatment, genes corresponding to structural constituents of ribosomes were not shown to be affected, but transcriptomic study revealed the up-regulation of translation initiation factors, such as EIF4A [63]. Several translation initiation factors have been characterized in response to priming in several species [79,96,98]. In fact, newly synthetized proteins have been shown in dynamic proteomic study during Arabidopsis seed imbibition [99] and by polysome activity assessment in sunflower even though protein turnover cannot be excluded as shown by the proteasome activation [100]. Moreover, genes involved in DNA methylation or acetylation were also affected in response to osmopriming, suggesting the activation of specific epigenetic modifications [97]. Such modifications may be responsible for stress tolerance of the emerging plant, according to Chen and Arora [101], who have suggested that priming induced a "memory" of priming-induced stress responses in the new plant [101]. Indeed, the epigenetic mechanisms are of great importance in plant stress memory, as reported for drought stress, for example (for review, see [102]).

"Priming memory" has been associated to oxidative stress defense [96]. They pointed out the importance of ROS-scavenging and the antioxidant defense system in improving germination and seedling growth of durum wheat under salt stress. Components of oxidative stress regulation were constantly associated with priming treatments and antioxidant priming induced germination improvement. H_2O_2 priming also has a beneficial effect on seed germination and salt tolerance associated to its signal role that triggers cell response rapidly when the emerging plant is exposed to environmental stress [103]. Enzymatic components such as CAT, PER, POX, GR, DHAR or peroxiredoxin were shown to be regulated at the transcription level in response to salt and PEG priming [63,96,104]. Non-enzymatic components represented by ASA and GSH were also affected by priming [58,66,96]. Other repair components, such as protein-L-isoaspartate O-methyltransferase (PIMT), protein disulfide isomerase-like 2–3 (PDIL2–3) and HSP7S can also operate to repair oxidized proteins [96,98]. Thus, the oxidative stress tolerance machinery set up during priming allowed the protection the seed cell from oxidative damage leading to successful germination even under stress conditions [105].

Molecular characterization of seed priming has also shown the involvement of proteins related to metabolic processes. Respiration is triggered by seed hydration and results in an increase in energy charge and then in cell activities and macromolecular biosynthesis during seed germination. It has been proposed that glycolysis-related enzymes increase in expression during the early phases of germination and the glycolysis represents the main provider of energy needs for germination [100,106–108]. Similarly to the early phases of germination, seed priming induces an increase in the abundance of proteins related to energy and carbohydrate metabolism [109,110] Additional studies on metabolites can bring new elements about the specificity of these pathways in priming induced germination performance.

Several genes and proteins belonging to cell wall, cytoskeleton or cell division classes have been characterized, such as xyloglucan endotransglucosylase/hydrolases (XTH), tubulin subunits or expansins, in response to different priming treatments, including hydro-, osmo- and PEG priming [63,79,111]. Such molecules seem to be essential for normal seed germination as they have been characterized during the imbibition phase II of germination sensu stricto [100]. They are likely associated with elongation and growth processes and more investigations are needed to understand their regulation in priming induced improvement.

Special attention was given to potential priming markers from molecular studies. Proteomic study on rice seeds, presented the α-amylase as an ideal candidate for germination performance [106]. On the other hand, according to the similarity of the pattern of expression at the gene and protein levels, Cheng et al. [97] proposed three proteins as better candidates for seed priming in rice, glucose-1-phosphate adenylyltransferase large subunit, aminotransferase and prolamin precursor. Oxidative stress [103] or protein carbonylation [112] were also proposed as good salt-priming biomarkers. In fact, there will be as many markers as the species studied or the priming protocol applied. It is therefore important to characterize the markers according to the species but also to the specific environmental conditions of field production. Such markers are very important for agriculture and will be even more so in the future due to global warming.

5. Conclusions

Primed seeds can be considered as high vigor seeds. The beneficial effects of priming are associated with numerous biochemical, cellular and molecular events including energy metabolism, sugar synthesis and synthesis of proteins, RNA and DNA, but epigenetic effects should also be considered. The priming technique is without doubt a good strategy to improve crop production, but its effects depend often on the cultivar and the initial vigor of the seeds. Moreover, large scale seed priming requires a precise control of seed imbibition in order to regulate the advancement of the physiological and cellular processes and avoid any radicle protrusion resulting in loss of desiccation tolerance, and then the inability of primed seeds to be dried.

This review was focused on conventional priming techniques such as hydro- and osmopriming, but an added benefit of this technique is that seeds can simultaneously be treated with various substances, such as hormone (gibberellins or kinetin), nutrients or H_2O_2, that can induce the antioxidant defense machinery. Bio-priming that integrates inoculation with specific microorganisms involved in the regulation of seed pathogens is also a promising technique in agriculture.

Finally, priming is now used in seed companies to improve seed germination and crop performance, but they have to take into consideration the risk of losing the beneficial effect of priming during drying and storage. Seed companies must thus control all steps of the process, from the priming treatment itself to seed drying and storage before sowing. We lack the omics—e.g., of research on the molecular/epigenetic processes involved during priming and in primed seeds after drying and during storage—necessary to determine the long-term impact of the priming treatment. Omics approaches are expected to deliver new markers of seed vigor or of the efficiency of priming that can be used in breeding programs.

Author Contributions: Conceptualization, F.C. and H.E.-M.-B.; Investigation, F.C., N.T.-Ö. and H.E.-M.-B.; Writing original draft, F.C., N.T.-Ö. and H.E.-M.-B.; Writing-review and editing, F.C. and H.E.-M.-B.; Supervision, F.C. All authors have read and agreed to the published version of the manuscript.

Funding: This research received no external funding.

Institutional Review Board Statement: Not applicable.

Informed Consent Statement: Not applicable.

Data Availability Statement: Not applicable.

Conflicts of Interest: The authors declare no conflict of interest.

References

1. Halmer, P. Methods to improve seed performance in the field. In *Handbook of Seed Physiology: Applications to Agriculture*; Benech-Arnold, R.L., Sanchez, R.A., Eds.; The Haworth Reference Press: New York, NY, USA; London, UK; Oxford, UK, 2004; pp. 125–166.
2. Corbineau, F.; Côme, D. Priming: A technique for improving seed quality. *Seed Test. Inter.* **2006**, *132*, 38–40.

3. Waqas, M.; Korres, N.E.; Khan, M.D.; Nizami, A.S.; Deeba, F.; Ali, I.; Hussain, H. Advances in the concept and methods of seed priming. In *Priming and Pretreatment of Seeds and Seedlings*; Hasanuzzaman, M., Fotopoulos, V., Eds.; Springer Nature Singapore Pte Ltd.: Singapore, 2019; pp. 11–41.
4. Kumar, P. A review on seed priming techniques in field crops. *Inter. J. Progress. Res. Sci. Eng.* **2020**, *1*, 86–91.
5. Garcia, D.; Zhao, S.; Arif, S.; Zhao, Y.; Ming, L.C.; Huang, D. Seed priming technology as a key strategy to increase crop plant production under adverse environmental conditions. *J. Agric. Hortic. Res.* **2022**, *5*, 27–35.
6. Harada, J.J. Seed maturation and control of germination. In *Cellular and Molecular Biology of Plant Seed Development*; Larkins, B., Vasil, I., Eds.; Kluwer Academic Publishers: Dordrecht, The Netherlands, 1997; pp. 545–592.
7. Al-Chaarani, G.R.; Gentzbittel, L.; Wedzony, M.; Sarrafi, A. Identification of QTLs for germination and seedling development in sunflower (*Helianthus annuus* L.). *Plant Sci.* **2005**, *169*, 221–227. [CrossRef]
8. Finch-Savage, W.E.; Clay, H.A.; Lynn, J.R.; Morris, K. Towards a genetic understanding of seed vigour in small-seeded crops using natural variation in *Brassica oleracea*. *Plant Sci.* **2010**, *179*, 582–589. [CrossRef]
9. Vandecasteele, C.; Teulat-Merah, B.; Morère-Le Paven, M.C.; Leprince, O.; Ly Vu, B.; Viau, L.; Ledroit, L.; Pelletier, S.; Payer, N.; Satour, P.; et al. Quantitative trait loci analysis reveals a correlation between the ratio of sucrose/raffinose family oligosaccharides and seed vigour in *Medicago truncatula*. *Plant Cell Environ.* **2011**, *34*, 1473–1487. [CrossRef]
10. Dias, P.; Brunel-Muguet, S.; Dürr, C.; Huguet, T.; Demilly, D.; Wagner, M.-H.; Teulat-Merah, B. QTL analysis of seed germination and pre-emergence growth at extreme temperatures in *Medicago truncatula*. *Theor. Appl. Genet.* **2011**, *122*, 429–444. [CrossRef]
11. McDonald, M.B. Seed quality assessment. *Seed Sci. Res.* **1998**, *8*, 265–275. [CrossRef]
12. McDonald, M.B. Seed Enhancements. In *Seed Science and Technology*; Copeland, L.O., McDonald, M.B., Eds.; Kluwer Academic Publishers: Dordrecht, The Netherlands, 2001; pp. 277–296.
13. Ligterink, W.; Joosen, R.V.L.; Hilhorst, H.W.M. Unravelling the complex trait of seed quality: Using natural variation through a combination of physiology, genetics and –omics technologies. *Seed Sci. Res.* **2012**, *22*, S45–S52. [CrossRef]
14. Nile, S.H.; Thiruvengadam, M.; Wang, Y.; Samynathan, R.; Shariati, M.A.; Rebezov, M.; Nile, A.; Sun, M.; Venkidasamy, B.; Xiao, J.; et al. Nano-priming as emerging seed priming technology for sustainable agriculture–recent developments and future perspectives. *J. Nanobiotechnol.* **2022**, *20*, 254. [CrossRef]
15. Lutts, S.; Benincasa, P.; Wojtyla, L.; Kubala, S.; Pace, R.; Lechowska, K.; Quinet, M.; Garnczarska, M. Seed priming: New comprehensive approaches for an old empirical technique. In *New Challenges in Seed Biology—Basic and Translational Research Driving Seed Technology*; Ajaujo, S., Balestrazzi, A., Eds.; IntechOpen: London, UK, 2016; pp. 1–46.
16. Guedes, A.C.; Cantliffe, D.J. Germination of lettuce seeds at high temperature after seed priming. *J. Am. Soc. Hortic. Sci.* **1980**, *105*, 777–781. [CrossRef]
17. Brocklehurst, P.A.; Dearman, J. Interactions between seed priming treatments and nine seedlots of carrot, celery and onion. I. Laboratory germination. *Ann. Appl. Biol.* **1983**, *102*, 577–584. [CrossRef]
18. Brocklehurst, P.A.; Dearman, J. Interactions between seed priming treatments and nine seedlots of carrot, celery and onion. II. Seedling emergence and plant growth. *Ann. Appl. Biol.* **1983**, *102*, 585–593. [CrossRef]
19. Corbineau, F.; Côme, D. Effects of priming on the germination of *Valerianella olitoria* seeds in relation with temperature and oxygen. *Acta Hortic.* **1990**, *267*, 191–197. [CrossRef]
20. Corbineau, F.; Picard, M.A.; Côme, D. Germinability of some vegetable seeds in relation to temperature and oxygen. In *Fourth International Worshop on Seeds. Basic and Applied Aspects of Seed Biology*; Côme, D., Corbineau, F., Eds.; ASFIS: Paris, France, 1993; Volume 3, pp. 1027–1032.
21. Corbineau, F.; Picard, M.A.; Côme, D. Germinability of leek seeds and its improvement by osmopriming. *Acta Hortic.* **1994**, *371*, 45–52. [CrossRef]
22. Capron, I.; Corbineau, F.; Dacher, F.; Job, C.; Côme, D.; Job, D. Sugar seed priming: Effects of priming conditions on germination, solubilization of 11-S globulin and accumulation of LEA proteins. *Seed Sci. Res.* **2000**, *10*, 243–254. [CrossRef]
23. Taylor, A.G.; Allen, P.S.; Bennett, M.A.; Bradford, K.J.; Burris, J.S.; Misra, M.K. Seed enhancements. *Seed Sci. Res.* **1998**, *8*, 245–256. [CrossRef]
24. Rowse, H.R. Methods of Priming Seeds. In *Priming and Pretreatment of Seeds and Seedlings*; Hasanuzzaman, M., Fotopoulos, V., Eds.; Springer: Singapore, 1991.
25. Warren, J.E.; Bennett, M.A. Seed hydration using the drum priming system. *Hortic. Sci.* **1997**, *31*, 1220–1221. [CrossRef]
26. Heydecker, W.; Higgins, J.; Gulliver, R.L. Accelerated germination by osmotic seed treatment. *Nature* **1973**, *246*, 42. [CrossRef]
27. Bray, C.M. Biochemical processes during osmopriming of seeds. In *Seed Development and Germination*; Kigel, J., Galili, G., Eds.; Marcel Dekker, Inc.: New York, NY, USA, 1995; pp. 767–789.
28. Rajjou, L.; Duval, M.; Gallardo, K.; Catusse, J.; Bally, J.; Job, C.; Job, D. Seed germination and vigor. *Annu. Rev. Plant Biol.* **2012**, *63*, 507–533. [CrossRef]
29. Kalal, P.R.; Jajoo, A. Priming with Zinc oxide nanoparticles improve germination and photosynthetic performance in wheat. *Plant Physiol. Biochem.* **2021**, *160*, 341–351. [CrossRef] [PubMed]
30. Reddy, P.P. (Ed.) Biopriming of seeds. In *Recent Advances in Crop Protection India*; Springer Science & Business Media: New Delhi, India, 2013; pp. 83–90.
31. Dearman, J.; Brocklehurst, P.A.; Drew, R.L.K. Effects of osmotic priming and ageing on the germination and emergence of carrot and leek seed. *Ann. Appl. Biol.* **1987**, *111*, 717–722. [CrossRef]

32. Özbingöl, N.; Corbineau, F.; Côme, D. Responses of tomato seeds to osmoconditioning as related to temperature and oxygen. *Seed Sci. Res.* **1998**, *8*, 377–384. [CrossRef]
33. Özbingöl, N.; Corbineau, F.; Groot, S.P.C.; Bino, R.J.; Côme, D. Activation of the cell cycle in tomato (*Lycopersicon esculentum* Mill.) seeds during osmoconditioning as related to temperature and oxygen. *Ann. Bot.* **1999**, *84*, 245–251. [CrossRef]
34. Côme, D.; Özbingöl, N.; Picard, M.A.; Corbineau, F. Beneficial effects of priming on seed quality. In *Progress in Seed Research. Conference Proceedings of the Second International Conference on Seed Science and Technology*; Taylor, A.G., Huang, X.L., Eds.; Agricultural Experimental Station: Geneva, Switzerland, 1998; pp. 257–263.
35. Smok, M.A.; Chojnowski, M.; Corbineau, F.; Côme, D. Effect of osmotic treatment on sunflower seed germination in relation with temperature and oxygen. In *Fourth International Workshop on Seeds: Basic and Applied Aspects of Seed Biology*; Côme, D., Corbineau, F., Eds.; ASFIS: Paris, France, 1993; Volume 3, pp. 1033–1038.
36. Chojnowski, M.; Corbineau, F.; Côme, D. Physiological and biochemical changes induced in sunflower seeds by osmopriming and subsequent drying, storage and aging. *Seed Sci. Res.* **1997**, *7*, 323–331. [CrossRef]
37. Bradford, K.J.; Côme, D.; Corbineau, F. Quantifying the oxygen sensitivity of seed germination using a population-based threshold model. *Seed Sci. Res.* **2007**, *17*, 33–43. [CrossRef]
38. Khan, A.A.; Peck, N.H.; Samimy, C. Seed osmoconditioning: Physiological and biochemical changes. *Isr. J. Bot.* **1980**, *29*, 133–144.
39. Dell'Aquila, A.; Taranto, G. Cell division and DNA-synthesis during osmopriming treatment and following germination in aged wheat embryos. *Seed Sci. Technol.* **1986**, *14*, 333–341.
40. Fujikura, Y.; Karssen, C.M. Effects of controlled deterioration and osmopriming on protein synthesis of cauliflower during early germination. *Seed Sci. Sci.* **1992**, *2*, 23–31. [CrossRef]
41. Van Pijlen, J.G.; Kraak, H.L.; Bino, R.J.; De Vos, C.H.R. Effects of ageing and osmopriming on germination characteristics and chromosome aberrations of tomato. *Seed Sci. Technol.* **1995**, *29*, 823–830.
42. Bailly, C.; Benamar, A.; Corbineau, F.; Côme, D. Free radical scavenging as affected by accelerated ageing and subsequent priming in sunflower seeds. *Physiol. Plant.* **1998**, *104*, 646–652. [CrossRef]
43. Georghiou, K.; Thanos, C.A.; Passam, H.C. Osmoconditioning as a means of counteracting the ageing of pepper seeds during high temperature storage. *Ann. Bot.* **1987**, *60*, 279–285. [CrossRef]
44. Côme, D.; Corbineau, F. *Dictionnaire de la Biologie des Semences et des Plantules*; Lavoisier: Paris, France, 2006; p. 226.
45. Karssen, C.M.; Haigh, A.; Van Der Toorn, P.; Weges, R. Physiological mechanisms involved in seed priming. In *Recent Advances in the Development and Germination of Seeds*; Taylorson, R.B., Ed.; Plenum Press: New York, NY, USA; London, UK, 1989; pp. 269–280.
46. Wu, L.; Huo, W.; Yao, D.; Li, M. Effects of solid matrix priming (SMP) and salt stress on broccoli and cauliflower seed germination and early seedling growth. *Sci. Hortic.* **2019**, *255*, 161–168. [CrossRef]
47. Nascimento, W.M.; Huber, D.J.; Cantliffe, D.J. Carrot seed germination and respiration at high temperature in response to seed maturity and priming. *Seed Sci. Technol.* **2013**, *41*, 164–169. [CrossRef]
48. Lanteri, S.; Kraak, H.L.; De Vos, C.H.R.; Bino, R.J. Effects of osmotic preconditioning on nuclear replication activity in seeds of pepper (*Capsicum annuum*). *Physiol. Plant.* **1993**, *89*, 433–440. [CrossRef]
49. Lanteri, S.; Portis, E.; Bergervoet, H.W.; Groot, S.P.C. Molecular markers for the priming of pepper seeds (*Capsicum annuum* L.). *J. Hortic. Sci. Biotechnol.* **2000**, *75*, 607–611. [CrossRef]
50. Sung, Y.; Cantliffe, D.J.; Nagata, R. Using a puncture test to identify the role of seed coverings on thermotolerant lettuce seed germination. *Am. Soc. Hortic. Sci.* **1998**, *123*, 1102–1110. [CrossRef]
51. Cantliffe, D.J.; Schuler, K.D.; Guedes, A.C. Overcoming seed dormancy in heat sensitive romaine lettuce by seed priming. *HortScience* **1981**, *16*, 196–198. [CrossRef]
52. Valdes, V.M.; Bradford, K.J. Effects of seed coating and osmopriming on the germination of lettuce seeds. *J. Am. Soc. Hortic. Sci.* **1987**, *112*, 153–156. [CrossRef]
53. Schwember, A.R.; Bradford, K.J. Drying rates following priming affect temperature sensitivity of germination and longevity of lettuce seeds. *Hortic. Sci.* **2005**, *40*, 778–781. [CrossRef]
54. Schwember, A.R.; Bradford, K.J. A genetic locus and gene expression pattern associated with he priming effect on lettuce seed germination at elevated temperature. *Plant Mol. Biol.* **2010**, *73*, 105–118. [CrossRef]
55. De Castro, R.D.; van Lammeren, A.A.M.; Groot, S.P.C.; Bino, R.J.; Hilhorst, H.W.M. Cell division and subsequent radicle protrusion in tomato seeds are inhibited by osmotic stress but DNA synthesis and formation of microtubular cytoskeleton are not. *Plant Physiol.* **2000**, *122*, 327–336. [CrossRef]
56. Cheng, Z.Y.; Bradford, K.J. Hydrothermal time analysis of tomato seed germination responses to priming treatments. *J. Exp. Bot.* **1999**, *50*, 89–99. [CrossRef]
57. Chen, K.; Fessehaie, A.; Arora, R. Dehydrin metabolism is altered during seed osmopriming and subsequent germination under chilling and desiccation in *Spinacia oleracea* L. cv. Bloomsdale: Possible role in stress tolerance. *Plant Sci.* **2012**, *183*, 27–36. [CrossRef]
58. Chen, K.; Arora, R. Dynamics of the antioxidant system during seed osmopriming, post-priming germination, and seedling establishment in spinach (*Spinacia oleracea*). *Plant Sci.* **2011**, *180*, 212–220. [CrossRef]
59. Khan, A.A.; Peck, N.H.; Taylor, A.G.; Samimy, C. Osmoconditioning of beet seed to improve emergence and yield in cold soils. *Agron. J.* **1983**, *75*, 788–794. [CrossRef]

60. Job, C.; Kersulec, A.; Ravasio, L.; Chareyre, S.; Pépin, R.; Job, D. The solubilization of the basic subunit of sugarbeet seed 11-S globulin during priming and early germination. *Seed Sci. Res.* **1997**, *7*, 225–243. [CrossRef]
61. Job, D.; Capron, I.; Job, C.; Dacher, F.; Corbineau, F.; Côme, D. Identification of germination-specific protein markers and their use in seed priming technology. In *Seed Biology: Advances and Applications*; Black, M., Bradford, K.J., Vazquez-Ramos, J., Eds.; CABI Publishing, CAB International: Oxon, UK, 2000; pp. 449–459.
62. Pace, R.; Benincasa, P.; Ghanem, M.E.; Quinet, M.; Lutts, S. Germination of untreated and primed seeds in rapeseed (*Brassica napus* var oleifera Del.) under salinity and low matric potential. *Exp. Agric.* **2012**, *48*, 238–251. [CrossRef]
63. Kubala, S.; Garnczarska, M.; Wojtyla, L.; Clippe, A.; Kosmala, A.; Zmienko, A.; Lutts, S.; Quinet, M. Deciphering priming-induced improvement of rapeseed (*Brassica napus* L.) germinaon through an integrated transcriptomic and proteomic approach. *Plant Sci.* **2015**, *231*, 94–113. [CrossRef]
64. Posmyk, M.; Corbineau, F.; Vinel, D.; Bailly, C.; Côme, D. Osmopriming reduces physiological and biochemical damage induced by chilling in soybean seeds. *Physiol. Plant.* **2001**, *111*, 473_4482. [CrossRef]
65. Posmyk, M.M.; Janas, K.M. Effects of seed hydropriming in presence of exogenous proline on chilling injury limitation in *Vigna radiata* L. seedlings. *Acta Physiol. Plant.* **2007**, *29*, 509–517. [CrossRef]
66. Sun, H.; Lin, L.; Wang, X.; Wu, S.; Wang, X. Ascorbate-glutathione cycle of mitochondria in osmoprimed soybean cotyledons in response to imbibitional chilling injury. *J. Plant Physiol.* **2011**, *168*, 226–232. [CrossRef] [PubMed]
67. Kaya, M.D.; Okc, G.; Atak, M.; Yakupa, C.O.K. Seed treatments to overcome salt and drought stress during germination in sunflower (*Helianthus annuun* L.). *Eur. J. Agron.* **2006**, *24*, 291–295. [CrossRef]
68. Bailly, C.; Benamar, A.; Corbineau, F.; Côme, D. Antioxidant system in sunflower (*Helianthus annuus* L.) seeds as affected by priming. *Seed Sci. Res.* **2000**, *10*, 35–42. [CrossRef]
69. Bailly, C.; Bogatek-Leszczynska, R.; Côme, D.; Corbineau, F. Changes in activities of antioxidant enzymes and lipoxygenase during growth of sunflower seedlings from seeds of different vigour. *Seed Sci. Res.* **2002**, *12*, 47–55. [CrossRef]
70. Gendreau, E.; Romaniello, S.; Barad, S.; Leymarie, J.; Benech-Arnold, R.; Corbineau, F. Regulation of cell cycle activity in the embryo of barley seeds during germination as related to grain hydration. *J. Exp. Bot.* **2008**, *59*, 203–212. [CrossRef]
71. Mondal, S.; Viji, P.; Bose, B. Role of seed hardening in rice variety Swarna (MTU 7029). *Res. J. Seed Sci.* **2011**, *4*, 157–165. [CrossRef]
72. Sen, A.; Puthur, J.T. Influence of different seed priming techniques on oxidative and antioxidative responses during the germination of *Oryza sativa* varieties. *Physiol. Mol. Biol. Plants* **2020**, *26*, 551–565. [CrossRef]
73. Farooq, M.; Wahid, A.; Kobayashi, N.; Fujita, D.; Basra, S.M. Plant drought stress: Effects, mechanisms and management. *Sustain. Agric.* **2009**, *29*, 153–188.
74. Li, X.; Zhang, L. SA and PEG-induced priming for water stress tolerance in rice seedling. In *Information Technology and Agricultural Engineering*; Zhu, E., Sambath, S., Eds.; Springer: Berlin/Heidelberg, Germany, 2012; pp. 881–887.
75. Zhang, F.; Yu, J.; Johnston, C.R.; Wang, Y.; Zhu, K.; Lu, F.; Lu, F.; Zhang, Z.; Zou, J. Seed priming with polyethylene glycol induces physiological changes in sorghum (*Sorghum bicolor* L. moench) seedlings under suboptimal soil moisture environments. *PLoS ONE* **2015**, *10*, e0140620. [CrossRef]
76. Farooq, M.; Basra, S.M.A.; Rehman, H.; Saleem, B.A. Seed priming enhances the performance of late sown wheat (*Triticum aestivum* L.) by improving chilling tolerance. *J. Agron. Crop Sci.* **2008**, *194*, 55–60. [CrossRef]
77. Farooq, M.; Basra, S.M.A.; Tabassum, R.; Afzal, I. Enhancing the performance of direct seeded fine rice by seed priming. *Plant Prod. Sci.* **2006**, *9*, 446–456. [CrossRef]
78. Farooq, M.; Irfan, M.; Aziz, T.; Ahmad, I.; Cheema, S.A. Seed priming with ascorbic acid improves drought resistance of wheat. *J. Agron. Crop Sci.* **2013**, *199*, 12–22. [CrossRef]
79. Gallardo, K.; Job, C.; Groot, S.P.C.; Puype, M.; Demol, H.; Vandekerckhove, J.; Job, D. Proteomic analysis of Arabidopsis seed germination and priming. *Plant Physiol.* **2001**, *126*, 835–848. [CrossRef]
80. Corbineau, F. Markers of seed quality: From present to future. *Seed Sci. Res.* **2012**, *22*, S61–S68. [CrossRef]
81. Corbineau, F.; Özbingöl, N.; Vinel, D.; Côme, D. Improvement of tomato seed germination by osmopriming as related to energy metabolism. In *Seed Biology: Advances and Applications*; Black, M., Bradford, K.J., Vazquez-Ramos, J., Eds.; CABI Publishing, CAB International: Wallingford, UK, 2000; pp. 467–476.
82. Lanteri, S.; Saracco, F.; Kraak, H.L.; Bino, R.J. The effects of priming on nuclear replication activity and germination of pepper (*Capsicum annuum*) and tomato (*Lycopersicon esculentum*) seeds. *Seed Sci. Res.* **1994**, *4*, 81–87. [CrossRef]
83. De Castro, R.D.; Zheng, X.; Bergervoet, J.H.W.; De Vos, C.H.; Bino, R.J. B-Tubulin accumulation and DNA replication in imbibing tomato seeds. *Plant Physiol.* **1995**, *109*, 499–504. [CrossRef]
84. Halpin-Ingham, B.; Sundstom, F.J. Pepper seed water content, germination response and respiration following priming treatment. *Seed Sci. Technol.* **1992**, *20*, 589–596.
85. Dahal, P.; Kim, N.-S.; Bradford, K.J. Respiration and germination rates of tomato seeds at suboptimal temperatures and reduced water potentials. *J. Exp. Bot.* **1996**, *47*, 941–947. [CrossRef]
86. Fu, J.R.; Lu, X.H.; Chen, R.Z.; Zhang, B.Z.; Liu, Z.S.; Li, Z.S.; Cai, D.Y. Osmoconditioning of peanut (*Arachis hypogea* L.) seeds with PEG to improve vigour and some biochemical activities. *Seed Sci. Technol.* **1988**, *16*, 197–212.
87. Khan, A.A. ACC-derived ethylene production, a sensitive test for seed vigour. *J. Am. Soc. Hortic. Sci.* **1994**, *119*, 1083–1090. [CrossRef]

88. Özbingöl, N. *Evénement Cellulaires et Métaboliques Associés à la Stimulation de la Germination des Grains de Tomate (Lycopersicon esculentum Mill) par un Traitement de Prégermination*; Thesis Université Pierre et Marie Curie: Paris, France, 1998; p. 118.
89. Bruggink, T.; Van der Toorn, P. Induction of desiccation tolerance in germinated seeds. *Seed Sci. Res.* **1995**, *5*, 1–4. [CrossRef]
90. Bailly, C. Active oxygen species and antioxidants in seed biology. *Seed Sci. Res.* **2004**, *14*, 93–107. [CrossRef]
91. Bailly, C. The signaling role of ROS in the regulation of seed germination and dormancy. *Biochem. J.* **2019**, *476*, 3019–3032. [CrossRef] [PubMed]
92. Bino, R.J.; De Vries, J.N.; Kraak, H.L.; Van Pijlen, J.G. Flow cytometric determination of nuclear DNA replication stages in tomato seeds during priming and germination. *Ann. Bot.* **1992**, *69*, 231–236. [CrossRef]
93. Saracco, F.; Bino, R.J.; Bergervoet, J.H.W.; Lanteri, S. Influence of priming-induced nuclear replication activity on storability of pepper (*Capsicum annuum* L.) seed. *Seed Sci. Res.* **1995**, *5*, 25–29. [CrossRef]
94. Redfearn, M.; Osborne, D.J. Effects of advancement on nucleic acids in sugarbeet (*Beta vulgaris*) seeds. *Seed Sci. Res.* **1997**, *7*, 261–267. [CrossRef]
95. Soeda, Y.; Konings, M.C.J.M.; Vorst, O.; van Houwelingen, A.M.M.L.; Stoopen, G.M.; Maliepaard, C.A.; Kodde, J.; Bino, R.J.; Groot, S.P.C.; van der Geest, A.H.M. Gene expression programs during *Brassica oleracea* seed maturation, osmopriming, and germination are indicators of progression of the germination process and the stress tolerance level. *Plant Physiol.* **2005**, *137*, 354–368. [CrossRef] [PubMed]
96. Fercha, A.; Capriotti, A.L.; Caruso, G.; Cavaliere, C.; Gherroucha, H.; Samperi, R.; Stampachiacchiere, S.; Lagana, A. Gel-free proteomics reveal potential biomarkers of priming-induced salt tolerance in durum wheat. *J. Proteom.* **2013**, *91*, 496–499. [CrossRef]
97. Cheng, J.; Wang, L.; Zeng, P.; He, Y.; Zhou, R.; Zhang, H.; Wang, Z. Identification of genes involved in rice seed priming in the early imbibition stage. *Plant Biol.* **2017**, *19*, 61–69. [CrossRef]
98. Yacoubi, R.; Job, C.; Belghazi, M.; Chaibi, W.; Job, D. Toward characterizing seed vigor in alfalfa through proteomic analysis of germination and priming. *J. Proteome Res.* **2011**, *10*, 3891–3903. [CrossRef]
99. Galland, M.; Huguet, R.; Arc, E.; Cueff, G.; Job, D.; Rajjou, L. Dynamic proteomics emphasizes the importance of selective mRNA translation and protein turnover during Arabidopsis seed germination. *Mol. Cell. Proteom.* **2014**, *13*, 252–268. [CrossRef] [PubMed]
100. Xia, Q.; Ponnaiah, M.; Cueff, G.; Rajjou, L.; Prodhomme, D.; Gibon, Y.; Bailly, C.; Corbineau, F.; Meimoun, P.; El-Maarouf-Bouteau, H. Integrating proteomics and enzymatic profiling to decipher seed metabolism affected by temperature in seed dormancy and germination. *Plant Sci.* **2018**, *269*, 118–125. [CrossRef] [PubMed]
101. Chen, K.; Arora, R. Priming memory invokes seed stress-tolerance. *Environ. Exp. Bot.* **2013**, *94*, 33–45. [CrossRef]
102. Godwin, J.; Farrona, S. Plant epigenetic stress memory induced by drought: A physiological and molecular perspective. *Methods Mol. Biol.* **2020**, *2093*, 243–259. [PubMed]
103. Ellouzi, H.; Sghayar, S.; Abdelly, C. H_2O_2 seed priming improves tolerance to salinity; drought and their combined effect more than mannitol in *Cakile maritima* when compared to *Eutrema salsugineum*. *J. Plant Physiol.* **2017**, *210*, 38–50. [CrossRef] [PubMed]
104. Li, F.; Wu, X.; Tsang, E.; Cutler, A.J. Transcriptional profiling of imbibed *Brassica napus* seed. *Genomics* **2005**, *86*, 718–730. [CrossRef]
105. El-Maarouf-Bouteau, H. The seed and the metabolism regulation. *Biology* **2022**, *11*, 168. [CrossRef]
106. Yang, P.; Li, X.; Wang, X.; Chen, H.; Chen, F.; Shen, S. Proteomic analysis of rice (*Oryza sativa*) seeds during germination. *Proteomics* **2007**, *7*, 3358–3368. [CrossRef]
107. Yu, Y.L.; Guo, G.F.; Lv, D.W.; Hu, Y.K.; Li, J.R.; Li, X.H.; Yan, Y. Transcriptome analysis during seed germination of elite Chinese bread wheat cultivar Jimai. *BMC Plant Biol.* **2014**, *14*, 1471–2229. [CrossRef]
108. Xu, H.H.; Liu, S.J.; Song, S.H.; Wang, R.X.; Wang, W.Q.; Song, S.Q. Proteomics analysis reveals distinct involvement of embryo and endosperm proteins during seed germination in dormant and non-dormant rice seeds. *Plant Physiol. Biochem.* **2016**, *103*, 219–242. [CrossRef] [PubMed]
109. Araujo, G.D.S.; Lopes, L.S.; Paula-Marinho, S.O.; Mesquita, R.O.; Nagano, C.S.; Vasconcelos, F.R.; de Carvalho, H.H.; Moura, A.A.A.N.; Marques, E.C.; Gomes-Filho, E. H2O2 priming induces proteomic reponses to defense against salt stress in maize. *Plant Mol. Biol.* **2021**, *106*, 33–48. [CrossRef] [PubMed]
110. Zhao, S.; Zou, H.; Jia, Y.; Pan, X.; Huang, D. Carrot (*Daucus carota* L.) Seed Germination was promoted by hydro-electro hybrid priming through regulating the accumulation of proteins involved in carbohydrate and protein metabolism. *Front. Plant Sci.* **2022**, *10*, 824439. [CrossRef] [PubMed]
111. Ferreira Ribas, A.; Volpi, E.; Silva, N.; Dos Santos, T.B.; Lima Abrantes, F.; Castilho Custódio, C.; Barbosa Machado-Neto, N.; Esteves Vieira, L.G. Regulation of α-expansins genes in *Arabidopsis thaliana* seeds during post-osmopriming germination. *Physiol. Mol. Biol. Plants* **2019**, *25*, 511–522. [CrossRef]
112. Boucelha, L.; Abrous-Belbachir, O.; Djebbar, R. Is protein carbonylation a biomarker of seed priming and ageing? *Funct. Plant Biol.* **2021**, *48*, 611–623. [CrossRef]

seeds MDPI

Review

Ascorbic Acid in Seeds, Priming and Beyond

Mattia Terzaghi [1] and Mario C. De Tullio [2,*]

[1] Department of Biosciences, Biotechnologies and Environment, University of Bari, 70126 Bari, Italy; mattia.terzaghi@uniba.it
[2] Department of Earth and Environmental Sciences, University of Bari, 70126 Bari, Italy
* Correspondence: mario.detullio@uniba.it

Abstract: Ascorbic acid (AsA) is mainly known as an antioxidant. However, if the peculiar features of the AsA system in the different stages of seed development and germinationare taken into consideration, it can be concluded that the function of AsA goes far beyond its antioxidant properties. The possible involvement of AsA in the regulation of hormone synthesis and in the epigenetic control of gene expression opens new directions to further research. In recent years, seed priming with AsA has been successfully used as a strategy to improve germination and plant productivity. Beneficial effects of seed AsA priming could be observed in several crop species, but the underlying molecular mechanism(s) are still unclear. The available evidence suggests that AsA priming induces a wide range of coordinated responses allowing primed seeds to overcome adverse environmental conditions.

Keywords: ascorbic acid; seed development; antioxidant priming; germination

Citation: Terzaghi, M.; De Tullio, M.C. Ascorbic Acid in Seeds, Priming and Beyond. *Seeds* **2023**, *2*, 421–435. https://doi.org/10.3390/seeds2040032

Academic Editors: José Antonio Hernández Cortés, Gregorio Barba-Espín and Pedro Diaz-Vivancos

Received: 9 January 2023
Revised: 26 October 2023
Accepted: 27 October 2023
Published: 1 November 2023

1. Introduction

The capability of producing seeds is one of the most successful features that appeared during plant evolution [1]. Seed-producing plants (spermatophytes) significantly improved the chances for their progeny to thrive even under adverse environmental conditions. Besides providing mechanical protection to the embryo and storing reserves that will be used for embryo development, seeds ensure embryo survival due to their capability of withstanding desiccation and reducing metabolic activity until suitable conditions for germination occur [2]. The implementation of the developmental program leading to seedling establishment requires extensive epigenetic and hormonal reprogramming [3,4]. Moreover, the transition from quiescence to active development is full of hidden perils and could lead to severe damaging of the embryo itself unless proper protection is prepared before seed dehydration or activated in parallel with the "awakening" of the germinating seed [5]. An increasing number of studies points at ascorbic acid (AsA) as a key player involved in all those processes. The dynamic regulation of the AsA system (including AsA production, utilization, and recycling of its oxidized forms) along the different stages of seed development, desiccation, after-ripening and germination supports the view that AsA availability varies as a function of the specific needs of each stage. Recent reports also show that seed priming using AsA treatment improves germination and plant performance, especially under stress conditions. The aim of the present contribution is providing a novel viewpoint on the different aspects of AsA function in seeds.

2. The AsA System

2.1. AsA Biosynthesis

All plants synthesize AsA following a biosynthetic route known as the D-mannose L-galactose (Smirnoff–Wheeler) pathway [6]. Additional entry points for AsA production may occur under specific conditions using the myo-inositol [7] and the galacturonate [8] pathways. The biosynthesis appears strictly controlled [9], with an AsA-dependent feedback mechanism inhibiting the expression of at least three key genes in the Smirnoff–Wheeler pathway, including the GDP-L-galactose phosphorylase (*GGP*) gene (known as *VTC2* in *Arabidopsis*) [10]. AsA biosynthesis is also regulated by different cues, including light and hormones [11]. Notably, AsA biosynthesis occurs in the cytosol, excepting for the terminal step, which takes place at the mitochondrial inner membrane and is catalyzed by the enzyme L-galactono-1,4-lactone dehydrogenase (L-GalLDH) using oxidized cytochrome *c* as the electron acceptor [12]. The acquisition of this mitochondrial step is a major difference between the way in which animals [13] and plants synthesize AsA and is likely to be an adaptation to a photoautotrophic lifestyle [14].

2.2. AsA Transport and Intracellular Distribution

Once synthesized in the mitochondria, AsA must be transported to all other organelles and cell compartments, where it is unevenly distributed. Immunocytochemical detection in *Arabidopsis* leaves shows higher AsA content in peroxisomes and the cytosol, a bit less in the nuclei, plastids and mitochondria, with vacuoles presenting the lowest AsA content [15]. There is also indication of a relatively small but significant apoplastic AsA pool possibly involved in redox sensing of extracellular cues [16]. The mechanism of AsA transport from the site of synthesis is not fully understood. It is known that AsA can cross lipid bilayers by means of passive diffusion [17], but it is generally assumed that the oxidized form dehydroascorbic acid (DHA, see below) diffuses more efficiently [16]. AsA transporters GLUT and SVTC have been detected and characterized in animal cells [18]. Much less is known about plant transporters. A mitochondrial ascorbic acid (MAT) transporter, apparently different from GLUT and SVTC ones, from rat liver and potato tuber has been partially characterized [19]. In addition, the *Arabidopsis* PHT4;4 protein, a member of the phosphate transporter 4 family, is responsible for chloride-dependent AsA transport into chloroplasts [20]. Additional transporters are likely to regulate AsA intracellular distribution.

2.3. AsA Utilization

Plant-specific AsA peroxidases (APX, EC 1.11.1.11) have been detected in the cytosol [21], mitochondria [22], peroxisomes [23], and chloroplasts, where two distinct APX forms (stromal and thylakoidal, respectively) occur [24]. The enzyme catalyzes hydrogen peroxide conversion to water and O_2, specifically using AsA as the electron donor. Rather than scavengers, APXs are responsible for the fine-tuning of hydrogen peroxide content [24], which is essential for signaling purposes [25]. Recent studies have also suggested that, at least in some plant species (including *Oryza sativa*, *Glycine max*, *Zea mays*, and species of the orchid genus *Oncidium*), cytosolic APX can also use glutathione (GSH) as an electron donor [26].

The blue-copper enyzme AsA oxidase (AO, EC 1.10.3.3) is another AsA-dependent enzyme whose physiological function has not been fully understood, although it is possibly involved in cell elongation [27] and in the establishment of root symbioses with arbuscular mycorrhizal fungi and rhizobacteria [28]. A search in the TAIR database (www.arabidopsis.org, accessed on 26 October 2023) retrieves four genes encoding putative AOs in the *Arabidopsis* genome: *AAO1* (At4g39830), *AAO2* (At5g21100), *AAO3* (At5g21105), and recently added *SRG1* (At1g17020).

Several members of the large class of 2-oxoglutarate-dependent dioxygenases (2-ODDs) utilize AsA in a complex reaction mechanism requiring, besides AsA and 2-oxoglutarate, also molecular oxygen and Fe^{2+} [29,30]. Different 2-ODDs catalyze a variety of reactions, including hydroxylation, epoxidation, and desaturation of specific substrates involved in the biosynthesis or the catabolism of plant hormones/regulators (ethylene, abscisic acid, gibberellins, auxin, salicylic acid) and a number of secondary metabolites. A very special subsection of 2-ODDs has been identified, involved in epigenetic mechanisms. These include TET hydroxylases, catalyzing the demethylation of methyl cytosine, and the Jumonji group of histone demethylases. TET hydroxylases have been fully characterized in animal cells, but indirect evidence suggests that plants also have them [31]. Histone demethylation activity catalyzed by specific 2-ODDs occurs during plant developmental processes and in response to stress conditions [32,33]. The actual amount of AsA consumed by each 2-ODD activity is hardly measurable. An indirect estimate has been obtained by measuring AsA content in *Arabidopsis* insertion lines in which different putative 2-ODDs had been inactivated. At least in some of those lines, AsA content was more than doubled [34], suggesting that some 2-ODDs could be responsible for a significant, if not massive, use of AsA.

Besides catalyzed utilization, AsA is known to react non-enzymically with reactive oxygen species (ROS) and with metal ions. According to available experimental evidence, AsA reacts preferentially with copper (II) and iron (III), rather than ROS [35]. On the other hand, it is unlikely that highly reactive radicals just freely travel around the cells, also considering the low permeability of polar molecules across biological membranes [36]. Therefore, AsA's direct interaction with ROS should not always be taken for granted.

2.4. Recycling of AsA Oxidized Forms

One-electron AsA oxidation produces monodehydroascorbate (MDHA, also known as ascorbate free radical). This free radical is short-lived and disproportionates to AsA and dehydroascorbic acid (DHA) [37]. Alternatively, MDHA can be reduced enzymatically by the NADH-dependent enzyme MDHA reductase (MDHAR, EC 1.6.5.4) located in the cytosol, mitochondria, chloroplasts, and peroxisomes [38]. Fully oxidized DHA, in turn, can be re-reduced to AsA by DHA reductase (DHAR, EC 1.8.5.1), using reduced glutathione (GSH) as the electron donor [39]. MDHAR and DHAR are considered "recycling enzymes", reconverting oxidized AsA forms back to AsA. However, recycling alone, in the absence of new biosynthesis, is apparently unable to keep up with the pace of AsA consumption [39,40]. This is also confirmed by the observation that seeds of the *Arabidopsis vtc2/vtc5* double mutant, which is not capable of de novo AsA synthesis, can start the germination process due to the recycling activity, but the seedlings are not viable in the absence of AsA supplementation [41]. It should also be considered that DHA reductase activity is performed by several proteins characterized by the C-X-X-C motif, including protein disulfide isomerase and glutaredoxins [42].

2.5. Possible DHA Signaling and Further Catabolism

In comparison with the huge amount of literature dealing with AsA biosynthesis and functions, relatively little attention has been given to the products of its degradation and catabolism. As mentioned above, the product of AsA oxidation is DHA, which is usually, but erroneously, represented as a three-carbonyl molecule, whereas it is preferentially in the dimeric form [43]. There is indication that DHA, rather than just the end product of AsA oxidation, should be considered a relevant signaling molecule [44–46], possibly in connection with its capability of reacting with thiols to form disulfide bonds [47]. A low AsA/DHA ratio in the apoplast is generally considered a proxy of stress [48].

If not recycled back to AsA (see above, Section 2.4), DHA undergoes irreversible degradation following two possible routes: either by hydrolysis yielding diketogulonic acid, or by oxidation, with the consequent production of oxalylthreonate, oxalate, and threonate [49–52]. Remarkably, different oxidizing agents produce different AsA degradation products, suggesting the possibility that they act as molecular signals [53]. Enzyme activities are involved in catabolism, but to our knowledge, the enzymes responsible have not been characterized yet. Interestingly, the inactivation of *Arabidopsis AtFAHD1a*, the gene encoding a fumarylacetoacetate hydrolase (FAH) domain-containing protein 1a, highly expressed in developing seeds, results in AsA, DHA, and threonic acid accumulation [54]. Recently, a new bacterial pathway of AsA degradation has been identified, involving novel enzymes and a FAH family member catalyzing the conversion of 2-keto-3-deoxy-L-lyxonate into 2-oxoglutarate (α-ketoglutarate) [55]. Hopefully, the full disclosure of the details of AsA catabolism in plants is quite close. This will possibly also help in understanding the physiological role of AsA degradation products, as in the case of the observed oxalyltransferase activity responsible for the transfer of oxalate groups from oxalylthreonate to carbohydrates [56].

3. Dynamic Regulation of the AsA System during Seed Development

During the different stages of their life, seeds apparently activate different items of the wide toolkit forming the AsA system [5]. The scheme in Figure 1 summarizes the main changes in the AsA system occurring during orthodox seed development, quiescence, and early germination stages. AsA content increases during seed development, then dramatically drops below detectability during dehydration. DHA content is very high during the cell elongation stage of seed development, then markedly decreases, although still remains detectable, in dry seeds [57,58]. No APX activity can be measured in dry seeds, but some DHAR and MDHAR activity is still observed. AsA oxidase is undetectable at all stages.

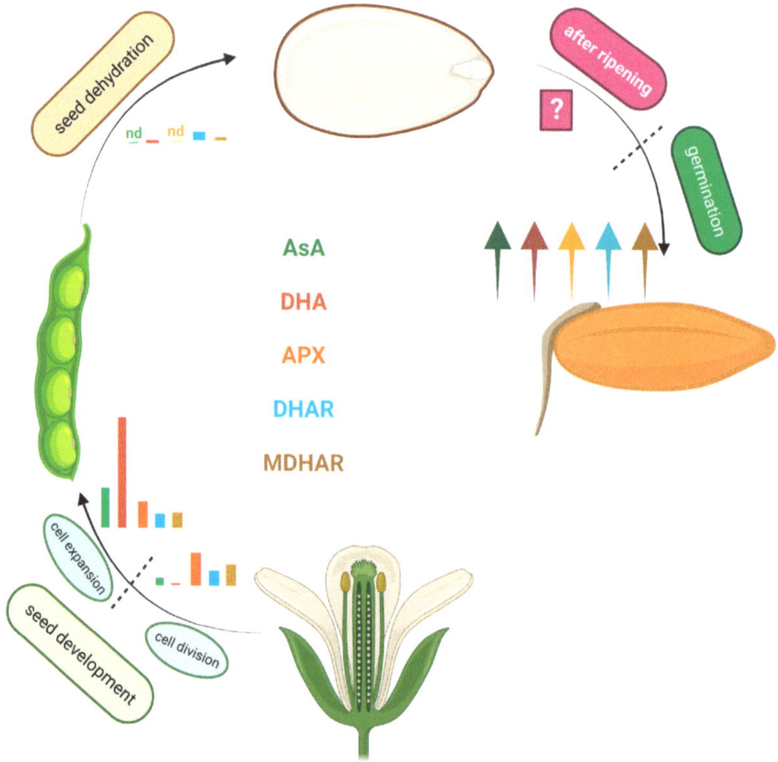

Figure 1. The ascorbic acid system in seeds. Relative ascorbate (AsA, green bars) and dehydroascorbate (DHA, red bars) content in seeds, and relative activity of the enzymes AsA peroxidase (APX, orange bars), DHA reductase (DHAR, cyan bars), and monodehydroascorbate reductase (MDHAR, brown bars) at different stages of seed life. AsA content and APX activity are not detectable (nd) in dry seeds. The question mark indicates the lack of detailed information on AsA dynamics during the after-ripening stage. The upward arrows at the stage of seed germination indicate a general increase of all the items considered. The scheme is based on the experimental data reported in [57,58].

It is worth noting that mature recalcitrant seeds, which undergo partial or no dehydration, retain AsA and APX activity [59]. Therefore, the absence of AsA and APX in orthodox seeds, in parallel with a low DHA supply and limited activity of the recycling enzymes, appears correlated with seed quiescence and longer viability.

The analysis of available *Arabidopsis* RNA-sequencing data [60] allows us to further characterize the changes occurring in seeds during the early stages of germination. By comparing the expression of the genes involved in the main AsA biosynthetic (Smirnoff-Wheeler) pathway (Figure 2), early activation of L-galactono-1,4-lactone dehydrogenase (L-GalLDH) occurs, in accordance with the observation that seeds become capable of converting L-GalL into AsA early during imbibition [5]. However, the expression of the main regulator of AsA biosynthesis, the *VTC2* gene encoding a GDP-L-galactose phosphorylase, appears rather low at this stage, suggesting that full-rate AsA biosynthesis is likely to occur only later in germination. Expression analysis of putative *DHAR* and *MDAR* genes (Figure 3) confirms the observation that recycling activities provide the small but essential AsA amount necessary to restart metabolic activity in germinating seeds, before de novo AsA biosynthesis becomes fully operational. Among putative *MDAR* genes, *MDAR1* and *MDAR4*, both encoding peroxisomal isoforms of the MDHA reductase enzyme, show early expression, paralleled by cytosolic *DHAR2*. It should, however, be considered that different plant species, or even different cultivars apparently manage the recycling of oxidized AsA forms in different ways. As an example, in germinating *Pisum sativum* cv. Alaska seeds, DHAR activity could not be detected [61], whereas it was 30-fold lower than MDAR activity in the cv. Lincoln [62].

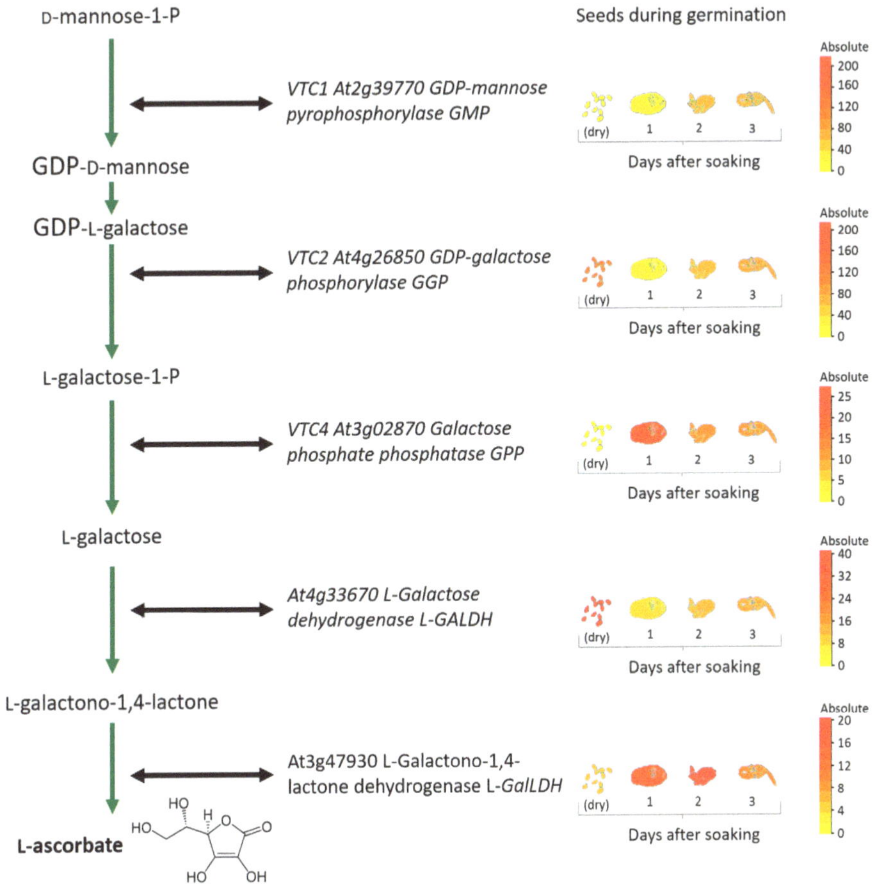

Figure 2. Expression of selected genes in the ascorbic acid (AsA) biosynthetic pathway (Smirnoff–Wheeler pathway). Data of RNA-sequencing experiments [60] retrieved from TAIR (The Arabidopsis Information Resource) website (www.arabidopsis.org, accessed on 26 October 2023).

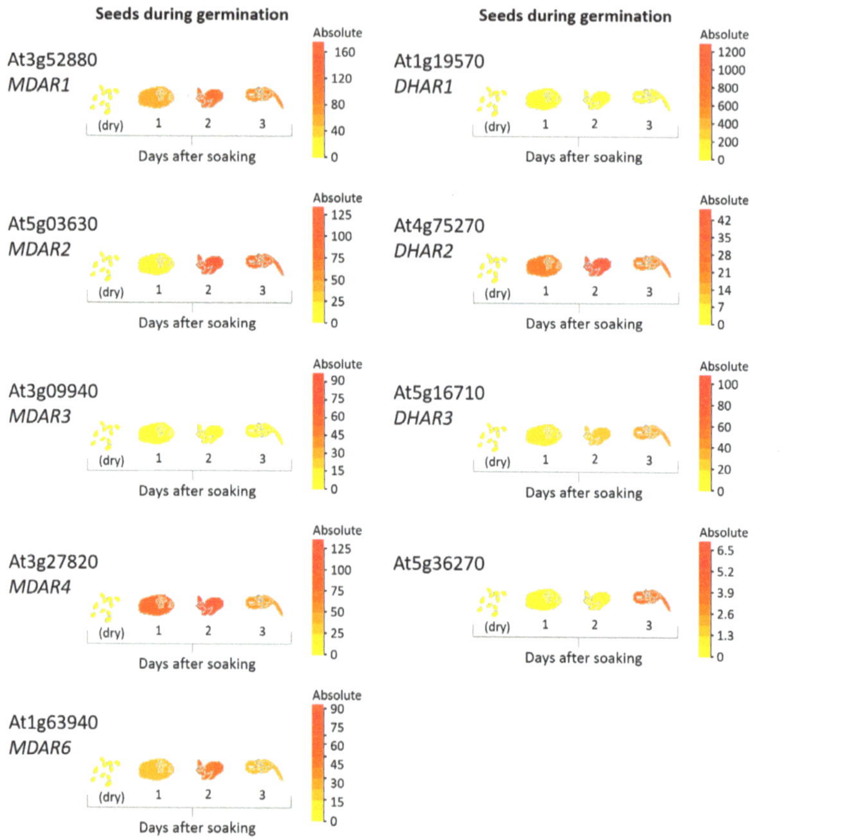

Figure 3. Expression of the *Arabidopsis* genes encoding putative monodehydroascorbate reductases (*MDAR*) and dehydroascorbate reductases (*DHAR*), respectively. At5g36270, previously considered a pseudogene because of the undetectability of the transcript, has been recently annotated as a *DHAR*. Data of RNA-sequencing experiments [60] retrieved from TAIR (The Arabidopsis Information Resource) website (www.arabidopsis.org, accessed on 26 October 2023).

4. AsA in Seed Dehydration, Dormancy, and Germination: To Be or Not to Be (There)?

The dynamic regulation of the AsA system outlined above shows that AsA and DHA are relevant players during seed development, and especially at the cell elongation stage, where DHA takes the lead. It should be considered that the entire process of development is controlled by the coordinated action of gibberellins and ABA [63]. AsA-dependent 2-ODDs (see Section 2.2 above) are involved in the biosynthesis of both hormones [64,65] and increased AsA content induces the expression of the ABA biosynthesis gene *9-cis-epoxycarotenoid dioxygenase NCED3* [66]. Due to its capability to oxidize protein thiols [47], DHA is also possibly responsible of the redox shift occurring in storage proteins, which at seed maturity are mostly in their oxidized (disulfide) state [67]. In the next stage (dehydrated seeds), AsA is totally absent, and only little DHA remains, suggesting not only that AsA is not required at this stage, but possibly that its presence could also negatively affect the dehydration step. Desiccation tolerance in living tissues is a complex process, not yet fully understood, but for sure it requires the interaction of several different players [68]. For orthodox seeds, desiccation precedes dormancy, which is the implementation of the safety mechanism(s) by which seeds do not germinate under "deceptive" favorable conditions occurring episodically, as in a short mild-weather period before full winter comes [2]. The establishment of dormancy is an ABA-dependent process regulated by the activity of NCED dioxygenases [65], requiring the presence of the co-substrate AsA [69], but the subsequent maintenance of dormancy apparently requires AsA removal. This possibility is strongly suggested by both old and recent findings. Besides the well-known presence of AsA and APX in recalcitrant (non-dormant) seeds [59], recent work by Gerna et al. [54] further supports this eventuality. As mentioned above regarding AsA catabolism (Section 2.5), the *Arabidopsis* mutant *Atfahd1a-1*, lacking fumarylacetoacetate hydrolase (FAH) activity, is apparently impaired in the AsA catabolic pathway and accumulates AsA, DHA, and threonic acid. Interestingly, the mutant shows shallower thermo-dormancy, together with increased seed longevity and a shift of the seed redox poise towards a reduced state. In the wild type, the *FAHD1a* gene is highly expressed in the embryo of fully mature and desiccated seeds. The total loss of AsA and the almost complete disappearance of DHA in dry seeds could be explained with the high activity of the FAHD1a enzyme (and possibly other uncharacterized AsA catabolic enzymes) at this stage. Oxalate, another product of AsA catabolism, is often accumulated in dry seeds, possibly regulating calcium uptake [70].

The fact that AsA content progressively decreases during seed maturation, so that dry seeds are devoid of AsA, is unlikely to be accidental. In a way, the absence of AsA in dry seeds is counterintuitive, especially in the general pervasive view that antioxidants, and AsA for one, always have a protective effect against any form of unfavorable environmental conditions, and seeds obviously need to be protected to increase their chances of survival. A possible explanation to this apparent contradiction could be the involvement of AsA in epigenetic mechanisms, namely in methyl-cytosine demethylation, in analogy with the well-characterized TET dioxygenases of animal cells [71]. The DNA methylation pattern markedly increases in developing embryos, keeps steady in quiescent seeds, and is then dramatically reversed with extensive demethylation at the very beginning of the germination process [72], in parallel with the recovery of AsA regeneration and de novo biosynthesis. It is tempting to hypothesize a causal relationship between these two events and a direct involvement of AsA in widespread DNA demethylation also in the cells of plant embryos, similarly to what is known to occur in human stem cells [73]. A second possibility to explain AsA absence in dry seeds is the well-known involvement of ROS in dormancy release [74,75]. Although known for years, only recently the mechanism of ROS-dependent dormancy alleviation has been better characterized at the cellular and molecular levels [76,77]. If AsA was stored in dry seeds, it could interfere with early ROS production that is key to starting germination, so its antioxidant action would be a burden rather than an advantage. Exciting new findings on dormancy and dormancy release have been reported in the last few years [78]. The identification of *DELAY OF GERMINATION-1* (*DOG1*) as a master regulator of dormancy opened new and unexpected directions in ongoing research on seed biology. The DOG1 protein appears involved in a complex signaling system involving ABA and possibly more, still uncharacterized, players [4,79]. Surprisingly, the DOG1 system is strictly connected to the expression of genes previously characterized for their involvement in the control of flowering time, including *FLOWERING LOCUS C* (*FLC*) and *FLOWERING LOCUS T* (*FT*) [80], the latter acting in two opposite configurations to regulate either flowering or dormancy release [81]. Histones associated with *DOG1* and *FLC* undergo extensive changes in their methylation patterns at the transition from dormancy to germination [82]. Indeed, both flowering and germination processes share the necessity of avoiding "false starts" that would jeopardize plant survival and life-cycle completion. It is worth mentioning that increased AsA content delays flowering and the expression of the *LEAFY* gene, which is expressed downstream of *FT* in the specification of floral organs [83].

Once germination starts, metabolic activity is fully recovered. Mitochondrial metabolism has a pivotal role at this stage [75,77]. Early ROS accumulation occurs in the mitochondria and is caused by the activation of the respiratory electron transport chain [84]. High expression of the gene encoding L-GalLDH, the mitochondria-located enzyme catalyzing the last step of AsA biosynthesis, occurs at this stage (Figure 2). As far as germination proceeds, ROS are found in the nuclei, where they possibly operate in the mechanism of chromatin decompaction, and later on in the peroxisomes [77]. It is conceivable that at this stage, AsA and APX become crucial to avoiding ROS overproduction. The expression of the six *Arabidopsis APX* genes during germination and in the seedlings is reported in Figure 4. *APX3*, encoding a peroxisomal APX, is expressed during germination. Early expression of the gene coding for the APX of chloroplast stroma is also observed, in a stage characterized by gibberellin-regulated rapid proplastid differentiation [85].

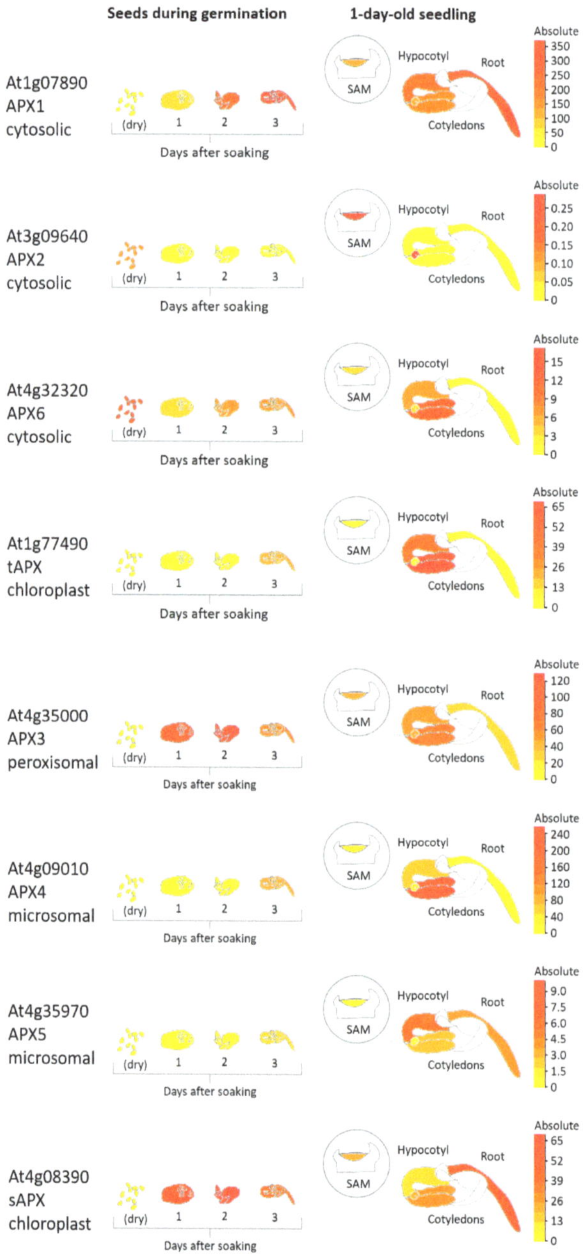

Figure 4. Expression of the *Arabidopsis* genes encoding putative ascorbate peroxidases (APX) in germinating seeds and in seedlings. Data of RNA-sequencing experiments [60] retrieved from TAIR (The Arabidopsis Information Resource) website (www.arabidopsis.org, accessed on 26 October 2023).

The control of ROS production by AsA and APX becomes more relevant under stress conditions. Inhibition of *Arabidopsis* seed germination in the presence of excess salt is regulated by a ROS-mediated signaling module connecting the transcription factor ABI4 with the NADPH oxidase gene *RbohD* and the AsA biosynthesis gene *VTC2* [86], confirming the centrality of AsA in the germination process. However, it should be considered that exogenous administration of AsA in excess inhibits rice seed germination, in a mechanism involving ABA and GA [87]. This means that endogenous AsA content must be carefully controlled to balance hormonal production and action.

5. Priming with AsA

Seed priming is an empirical practice known since antiquity [88] that proved to be a powerful tool in improving seed germination and plant performance, especially under adverse environmental conditions [89–91]. The term antioxidant priming is used to indicate seed treatment with different molecules potentially reacting with ROS, including AsA. Although a simple internet search using together the words "seed", "priming", and "ascorbic" retrieves a large number of papers whose title and abstract suggest a clear-cut positive effect of AsA priming on many different parameters of plant productivity, a closer look into some of those articles reveals a lack of proper controls and other flaws in the experimental design, thus making the data of those reports difficult to interpret. Even not considering those flawed papers, a substantial amount of sound experimental data confirm that priming seeds with AsA at different concentrations is beneficial to plant growth, development, and productivity. A very short list of papers comparing the effects of AsA priming on mean emergence time (MET) is presented in Table 1.

Table 1. Mean emergence time (MET) in seeds of crop species subjected to ascorbic acid (AsA) priming. ns: not significant.

AsA Concentration	MET (Days Ahead of Controls)	Species	Ref.
10 mg/L	0.36 (ns)	*Oryza sativa*	[92]
50 mg/L	2	*Triticum aestivum*	[93]
50 mg/L	1.1	*Triticum aestivum*	[94]
40 mg/L	0.68	*Zea mays*	[95]
2 mM	0.92	*Triticum aestivum*	[96]

Plant materials (cultivars) and treatment conditions vary in the different experiments reported in Table 1, but a tendency of early seedling emergence in AsA-primed seeds is generally observed. AsA priming also increases, although to a different extent, germination percentage, germination uniformity, and vegetative and reproductive growth in a variety of model and non-model plant species, or even improves nutrient profiles in seeds harvested from plants originally subjected to AsA priming at the seed stage [97]. However, the most convincing results are obtained when the germination of AsA-primed seeds takes place under stress conditions. As an example, the germination percentage of wheat plants in the presence of 200 mM of NaCl is 55 ± 6.5 in unprimed controls, but increases to 73 ± 6.6 when the seeds are primed with a 150 mg/L AsA solution [98]. Unfortunately, not many attempts have been made to explain the molecular mechanisms underlying the beneficial effects of AsA priming. Most studies simply advance the hypothesis that AsA priming improves general antioxidant defenses [99–102]. Only a few studies tried to go deeper into detail. An accurate analysis in artificially aged oat seeds has shown a repair effect of AsA and GSH priming on damaged mitochondria [103]. An interesting study analyzing the effect of AsA priming (0.5 mmol/L for 12 h) on wheat seed proteome, with or without NaCl (250 mmol/L solution), has shown altered expression of 167 proteins, the majority of which were under-regulated [104]. Most interestingly, AsA priming impacted negatively defense-related proteins, including antioxidants superoxide dismutase and AsA peroxidases. Such proteins were less represented in primed seeds (in both embryo and surrounding tissues) as compared to controls, and even less in primed seeds treated with NaCl. The presence of the AsA biosynthetic enzyme GDP-mannose-3,5-epimerase was also lower in primed seeds. This is in clear contrast with the claim that AsA priming is effective because it improves antioxidant defenses. Proteome data suggest that AsA priming induces a complex response that upregulates proteins involved in metabolism/energy and downregulates defense-related proteins, a picture that cannot be explained with simplistic considerations and unsupported assumptions based on AsA antioxidant properties. The effect of AsA priming on the after-ripening stage remains to be investigated.

6. What's Next?

Environmental stresses caused by climate change pose new constraints to plant growth and productivity. Investigating the mechanisms regulating seed germination under stress conditions will help in selecting tolerant genotypes of crop plants able to grow and reproduce in unfavorable environments. Several lines of reasoning point at a central role of AsA as a multi-level regulator of seed developmental and germination processes, making this peculiar molecule a promising target for further investigations. Unfortunately, AsA suffers the prejudice of being essentially categorized as an antioxidant, which is probably only a small part of its complex biochemical function. As discussed in the previous sections, there are sufficient indications that AsA is required for hormone synthesis and epigenetic regulation of gene expression (DNA and histone demethylation), although the details of this functional AsA dependency are still little known. The positive effects of seed priming with AsA, especially under unfavorable environmental conditions, offers an outstanding opportunity to deepen our understanding of the mechanisms controlling seed dormancy and germination. As discussed by Munns and Gilliham [105], plants growing in saline soils (or other stressful situations) pay their dues in terms of energy costs, at the expenses of their progeny (which, from an agricultural point of view, means at the expenses of plant productivity). Studies in the animal field suggest that epigenetic mechanisms can bring selective advantages, so that an organism endangered by environmental conditions invests in the improvement of progeny fitness [106]. AsA priming apparently goes in this direction, as suggested by the observation that corn plants derived from AsA-primed seeds produce seeds with improved vigor and protein content in the next generation [97], which is good both for the plants and for the heterotrophs consuming those seeds. It is especially interesting that a surprising shift occurs in the proteome of AsA-primed wheat seeds germinating in the presence of NaCl, with an increase of energy and metabolism-related proteins and a decrease in defense-related proteins [104]. The idea of a trade-off between energy and defense (immunity) is at the basis of ecological immunology, an area of research investigating how individual defense responses are integrated in the framework of environmental cues [107]. Ecological epigenetics is also a field in rapid expansion that will possibly provide answers to many questions currently under debate [108]. For sure, in order to untangle the complex interrelation of cues involved in the crucial process of germination and understand the actual contribution of AsA within this process, we need to think out of the box and explore new directions. This will be a challenge for the years to come.

Author Contributions: Conceptualization, original draft preparation, M.C.D.T. Review and editing, preparation of Figures 2–4, M.T. All authors have read and agreed to the published version of the manuscript.

Funding: This research was partly supported by the Italian Ministry of University and Research, PRIN grant 2022LLATJH.

Data Availability Statement: No new data were produced.

Acknowledgments: Original figures were created with BioRender.com.

Conflicts of Interest: The authors declare no conflict of interest.

References

1. Linkies, A.; Graeber, K.; Knight, C.; Leubner-Metzger, G. The evolution of seeds. *New Phytol.* **2010**, *186*, 817–831. [CrossRef] [PubMed]
2. Finch-Savage, W.E.; Leubner-Metzger, G. Seed dormancy and the control of germination. *New Phytol.* **2006**, *171*, 501–523. [CrossRef] [PubMed]
3. Kravets, O.P.; Sokolova, D.O. Epigenetic mechanisms regulating seed germination rate. *Cytol. Genet.* **2017**, *51*, 346–351. [CrossRef]
4. Smolikova, G.; Strygina, K.; Krylova, E.; Leonova, T.; Frolov, A.; Khlestkina, E.; Medvedev, S. Transition from Seeds to Seedlings: Hormonal and Epigenetic Aspects. *Plants* **2021**, *10*, 1884. [CrossRef]
5. De Tullio, M.C.; Arrigoni, O. The ascorbic acid system in seeds: To protect and to serve. *Seed Sci. Res.* **2003**, *13*, 249–260. [CrossRef]
6. Wheeler, G.L.; Jones, M.A.; Smirnoff, N. The biosynthetic pathway of vitamin C in higher plants. *Nature* **1998**, *393*, 365–369. [CrossRef]
7. Lorence, A.; Chevone, B.I.; Mendes, P.; Nessler, C.L. myo-inositol oxygenase offers a possible entry point into plant ascorbate biosynthesis. *Plant Physiol.* **2004**, *134*, 1200–1205. [CrossRef]
8. Agius, F.; González-Lamothe, R.; Caballero, J.L.; Muñoz-Blanco, J.; Botella, M.A.; Valpuesta, V. Engineering increased vitamin C levels in plants by overexpression of a D-galacturonic acid reductase. *Nat. Biotechnol.* **2003**, *21*, 177–181. [CrossRef]
9. Terzaghi, M.; De Tullio, M.C. The perils of planning strategies to increase vitamin C content in plants: Beyond the hype. *Front. Plant Sci.* **2022**, *13*, 1096549. [CrossRef]
10. Fenech, M.; Amorim-Silva, V.; Esteban Del Valle, A.; Arnaud, D.; Ruiz-Lopez, N.; Castillo, A.G.; Smirnoff, N.; Botella, M.A. The role of GDP-l-galactose phosphorylase in the control of ascorbate biosynthesis. *Plant Physiol.* **2021**, *185*, 1574–1594. [CrossRef]
11. Bulley, S.; Laing, W. The regulation of ascorbate biosynthesis. *Curr. Opin. Plant Biol.* **2016**, *33*, 15–22. [CrossRef] [PubMed]
12. Leferink, N.G.; van den Berg, W.A.; van Berkel, W.J. l-Galactono-gamma-lactone dehydrogenase from Arabidopsis thaliana, a flavoprotein involved in vitamin C biosynthesis. *Febs. J.* **2008**, *275*, 713–726. [CrossRef] [PubMed]
13. Duque, P.; Vieira, C.P.; Vieira, J. Advances in Novel Animal Vitamin C Biosynthesis Pathways and the Role of Prokaryote-Based Inferences to Understand Their Origin. *Genes* **2022**, *13*, 1917. [CrossRef] [PubMed]
14. Wheeler, G.; Ishikawa, T.; Pornsaksit, V.; Smirnoff, N. Evolution of alternative biosynthetic pathways for vitamin C following plastid acquisition in photosynthetic eukaryotes. *eLife* **2015**, *4*, e06369. [CrossRef] [PubMed]
15. Zechmann, B.; Stumpe, M.; Mauch, F. Immunocytochemical determination of the subcellular distribution of ascorbate in plants. *Planta* **2011**, *233*, 1–12. [CrossRef]
16. Horemans, N.; Foyer, C.H.; Asard, H. Transport and action of ascorbate at the plant plasma membrane. *Trends Plant Sci.* **2000**, *5*, 263–267. [CrossRef]
17. Łukawski, M.; Dałek, P.; Witkiewicz, W.; Przybyło, M.; Langner, M. Experimental evidence and physiological significance of the ascorbate passive diffusion through the lipid bilayer. *Chem. Phys. Lipids* **2020**, *232*, 104950. [CrossRef]
18. Corti, A.; Casini, A.F.; Pompella, A. Cellular pathways for transport and efflux of ascorbate and dehydroascorbate. *Arch Biochem. Biophys.* **2010**, *500*, 107–115. [CrossRef]
19. Scalera, V.; Giangregorio, N.; De Leonardis, S.; Console, L.; Carulli, E.S.; Tonazzi, A. Characterization of a Novel Mitochondrial Ascorbate Transporter From Rat Liver and Potato Mitochondria. *Front. Mol. Biosci.* **2018**, *5*, 58. [CrossRef]
20. Miyaji, T.; Kuromori, T.; Takeuchi, Y.; Yamaji, N.; Yokosho, K.; Shimazawa, A.; Sugimoto, E.; Omote, H.; Ma, J.F.; Shinozaki, K.; et al. AtPHT4;4 is a chloroplast-localized ascorbate transporter in Arabidopsis. *Nat. Commun.* **2015**, *6*, 5928. [CrossRef]
21. Maruta, T.; Inoue, T.; Noshi, M.; Tamoi, M.; Yabuta, Y.; Yoshimura, K.; Ishikawa, T.; Shigeoka, S. Cytosolic ascorbate peroxidase 1 protects organelles against oxidative stress by wounding- and jasmonate-induced H(2)O(2) in Arabidopsis plants. *Biochim. Biophys. Acta.* **2012**, *1820*, 1901–1907. [CrossRef]
22. Yin, B.; Zhang, J.; Liu, Y.; Pan, X.; Zhao, Z.; Li, H.; Zhang, C.; Li, C.; Du, X.; Li, Y.; et al. PtomtAPX, a mitochondrial ascorbate peroxidase, plays an important role in maintaining the redox balance of Populus tomentosa Carr. *Sci. Rep.* **2019**, *9*, 19541. [CrossRef]
23. Narendra, S.; Venkataramani, S.; Shen, G.; Wang, J.; Pasapula, V.; Lin, Y.; Kornyeyev, D.; Holaday, A.S.; Zhang, H. The Arabidopsis ascorbate peroxidase 3 is a peroxisomal membrane-bound antioxidant enzyme and is dispensable for Arabidopsis growth and development. *J. Exp. Bot.* **2006**, *57*, 3033–3042. [CrossRef]
24. Maruta, T.; Sawa, Y.; Shigeoka, S.; Ishikawa, T. Diversity and Evolution of Ascorbate Peroxidase Functions in Chloroplasts: More Than Just a Classical Antioxidant Enzyme? *Plant Cell Physiol.* **2016**, *57*, 1377–1386. [CrossRef]
25. Smirnoff, N.; Arnaud, D. Hydrogen peroxide metabolism and functions in plants. *New Phytol.* **2019**, *221*, 1197–1214. [CrossRef]
26. Chin, D.-C.; Senthil Kumar, R.; Suen, C.-S.; Chien, C.-Y.; Hwang, M.-J.; Hsu, C.-H.; Xuhan, X.; Lai, Z.X.; Yeh, K.-W. Plant Cytosolic Ascorbate Peroxidase with Dual Catalytic Activity Modulates Abiotic Stress Tolerances. *iScience* **2019**, *16*, 31–49. [CrossRef]
27. Stevens, R.; Truffault, V.; Baldet, P.; Gautier, H. Ascorbate Oxidase in Plant Growth, Development, and Stress Tolerance. In *Ascorbic Acid in Plant Growth, Development and Stress Tolerance*; Hossain, M.A., Munné-Bosch, S., Burritt, D.J., Diaz-Vivancos, P., Fujita, M., Lorence, A., Eds.; Springer International Publishing: Cham, Switzerland, 2017; pp. 273–295. [CrossRef]

28. Balestrini, R.; Ott, T.; Güther, M.; Bonfante, P.; Udvardi, M.K.; De Tullio, M.C. Ascorbate oxidase: The unexpected involvement of a 'wasteful enzyme' in the symbioses with nitrogen-fixing bacteria and arbuscular mycorrhizal fungi. *Plant Physiol. Biochem.* **2012**, *59*, 71–79. [CrossRef]
29. Hagel, J.M.; Facchini, P.J. Expanding the roles for 2-oxoglutarate-dependent oxygenases in plant metabolism. *Nat. Prod. Rep.* **2018**, *35*, 721–734. [CrossRef]
30. Kawai, Y.; Ono, E.; Mizutani, M. Evolution and diversity of the 2–oxoglutarate-dependent dioxygenase superfamily in plants. *Plant J.* **2014**, *78*, 328–343. [CrossRef]
31. Lucibelli, F.; Valoroso, M.C.; Aceto, S. Plant DNA Methylation: An Epigenetic Mark in Development, Environmental Interactions, and Evolution. *Int. J. Mol. Sci.* **2022**, *23*, 8299. [CrossRef]
32. Chen, X.; Hu, Y.; Zhou, D.X. Epigenetic gene regulation by plant Jumonji group of histone demethylase. *Biochim. Biophys. Acta.* **2011**, *1809*, 421–426. [CrossRef] [PubMed]
33. Chen, B.; Ali, S.; Zhang, X.; Zhang, Y.; Wang, M.; Zhang, Q.; Xie, L. Genome-wide identification, classification, and expression analysis of the JmjC domain-containing histone demethylase gene family in birch. *BMC Genom.* **2021**, *22*, 772. [CrossRef] [PubMed]
34. Mahmood, A.M.; Dunwell, J.M. 2-oxoglutarate-dependent dioxygenases: A renaissance in attention for ascorbic acid in plants. *PLoS ONE* **2020**, *15*, e0242833. [CrossRef]
35. Shen, J.; Griffiths, P.T.; Campbell, S.J.; Utinger, B.; Kalberer, M.; Paulson, S.E. Ascorbate oxidation by iron, copper and reactive oxygen species: Review, model development, and derivation of key rate constants. *Sci. Rep.* **2021**, *11*, 7417. [CrossRef]
36. Möller, M.N.; Cuevasanta, E.; Orrico, F.; Lopez, A.C.; Thomson, L.; Denicola, A. Diffusion and Transport of Reactive Species Across Cell Membranes. In *Bioactive Lipids in Health and Disease*; Trostchansky, A., Rubbo, H., Eds.; Springer International Publishing: Cham, Switzerland, 2019; pp. 3–19. [CrossRef]
37. Arrigoni, O. Ascorbate system in plant development. *J. Bioenerg. Biomembr.* **1994**, *26*, 407–419. [CrossRef]
38. Leterrier, M.; Cagnac, O. Function of the Various MDAR Isoforms in Higher Plants. In *Antioxidants and Antioxidant Enzymes in Higher Plants*; Gupta, D.K., Palma, J.M., Corpas, F.J., Eds.; Springer International Publishing: Cham, Switzerland, 2018; pp. 83–94. [CrossRef]
39. Ding, H.; Wang, B.; Han, Y.; Li, S. The pivotal function of dehydroascorbate reductase in glutathione homeostasis in plants. *J. Exp. Bot.* **2020**, *71*, 3405–3416. [CrossRef]
40. De Tullio, M.C.; De Gara, L.; Paciolla, C.; Arrigoni, O. Dehydroascorbate-reducing proteins in maize are induced by the ascorbate biosynthesis inhibitor lycorine. *Plant Physiol. Biochem.* **1998**, *36*, 433–440. [CrossRef]
41. Dowdle, J.; Ishikawa, T.; Gatzek, S.; Rolinski, S.; Smirnoff, N. Two genes in Arabidopsis thaliana encoding GDP-L-galactose phosphorylase are required for ascorbate biosynthesis and seedling viability. *Plant J.* **2007**, *52*, 673–689. [CrossRef]
42. De Tullio, M.C.; Paciolla, C.; Arrigoni, O. Identification and Analysis of Proteins Sharing Dehydroascorbate Reductase Activity. *Biol. Plant.* **2002**, *45*, 145–147. [CrossRef]
43. Njus, D.; Kelley, P.M.; Tu, Y.J.; Schlegel, H.B. Ascorbic acid: The chemistry underlying its antioxidant properties. *Free Radic. Biol. Med.* **2020**, *159*, 37–43. [CrossRef]
44. Fotopoulos, V.; De Tullio, M.C.; Barnes, J.; Kanellis, A.K. Altered stomatal dynamics in ascorbate oxidase over-expressing tobacco plants suggest a role for dehydroascorbate signalling. *J. Exp. Bot.* **2008**, *59*, 729–737. [CrossRef] [PubMed]
45. Chavan, S.N.; De Kesel, J.; Desmedt, W.; Degroote, E.; Singh, R.R.; Nguyen, G.T.; Demeestere, K.; De Meyer, T.; Kyndt, T. Dehydroascorbate induces plant resistance in rice against root-knot nematode Meloidogyne graminicola. *Mol. Plant Pathol.* **2022**, *23*, 1303–1319. [CrossRef] [PubMed]
46. Fiorani, M.; Azzolini, C.; Guidarelli, A.; Cerioni, L.; Cantoni, O. A novel biological role of dehydroascorbic acid: Inhibition of Na+-dependent transport of ascorbic acid. *Pharmacol. Res.* **2014**, *84*, 12–17. [CrossRef]
47. Saaranen, M.J.; Karala, A.R.; Lappi, A.K.; Ruddock, L.W. The role of dehydroascorbate in disulfide bond formation. *Antioxid. Redox Signal.* **2010**, *12*, 15–25. [CrossRef]
48. Pignocchi, C.; Foyer, C.H. Apoplastic ascorbate metabolism and its role in the regulation of cell signalling. *Curr. Opin. Plant Biol.* **2003**, *6*, 379–389. [CrossRef]
49. Green, M.A.; Fry, S.C. Vitamin C degradation in plant cells via enzymatic hydrolysis of 4-O-oxalyl-L-threonate. *Nature* **2005**, *433*, 83–87. [CrossRef]
50. Parsons, H.T.; Yasmin, T.; Fry, S.C. Alternative pathways of dehydroascorbic acid degradation in vitro and in plant cell cultures: Novel insights into vitamin C catabolism. *Biochem. J.* **2011**, *440*, 375–383. [CrossRef]
51. Parsons, H.T.; Fry, S.C. Oxidation of dehydroascorbic acid and 2,3-diketogulonate under plant apoplastic conditions. *Phytochemistry* **2012**, *75*, 41–49. [CrossRef]
52. Truffault, V.; Fry, S.C.; Stevens, R.G.; Gautier, H. Ascorbate degradation in tomato leads to accumulation of oxalate, threonate and oxalyl threonate. *Plant J.* **2017**, *89*, 996–1008. [CrossRef]
53. Dewhirst, R.A.; Fry, S.C. The oxidation of dehydroascorbic acid and 2,3-diketogulonate by distinct reactive oxygen species. *Biochem. J.* **2018**, *475*, 3451–3470. [CrossRef]
54. Gerna, D.; Arc, E.; Holzknecht, M.; Roach, T.; Jansen-Dürr, P.; Weiss, A.K.H.; Kranner, I. AtFAHD1a: A New Player Influencing Seed Longevity and Dormancy in Arabidopsis? *Int. J. Mol. Sci.* **2021**, *22*, 2997. [CrossRef]

55. Stack, T.M.M.; Morrison, K.N.; Dettmer, T.M.; Wille, B.; Kim, C.; Joyce, R.; Jermain, M.; Naing, Y.T.; Bhatti, K.; Francisco, B.S.; et al. Characterization of an l-Ascorbate Catabolic Pathway with Unprecedented Enzymatic Transformations. *J. Am. Chem. Soc.* **2020**, *142*, 1657–1661. [CrossRef] [PubMed]
56. Dewhirst, R.A.; Fry, S.C. Oxalyltransferase, a plant cell-wall acyltransferase activity, transfers oxalate groups from ascorbate metabolites to carbohydrates. *Plant J.* **2018**, *95*, 743–757. [CrossRef] [PubMed]
57. Arrigoni, O.; De Gara, L.; Tommasi, F.; Liso, R. Changes in the Ascorbate System during Seed Development of *Vicia faba* L. *Plant Physiol.* **1992**, *99*, 235–238. [CrossRef] [PubMed]
58. De Gara, L.; de Pinto, M.C.; Moliterni, V.M.C.; D'Egidio, M.G. Redox regulation and storage processes during maturation in kernels of Triticum durum. *J. Exp. Bot.* **2003**, *54*, 249–258. [CrossRef]
59. Tommasi, F.; Paciolla, C.; Arrigoni, O. The ascorbate system in recalcitrant and orthodox seeds. *Physiol. Plant.* **1999**, *105*, 193–198. [CrossRef]
60. Klepikova, A.V.; Kasianov, A.S.; Gerasimov, E.S.; Logacheva, M.D.; Penin, A.A. A high resolution map of the Arabidopsis thaliana developmental transcriptome based on RNA-seq profiling. *Plant J.* **2016**, *88*, 1058–1070. [CrossRef]
61. Barba-Espin, G.; Diaz-Vivancos, P.; Clemente-Moreno, M.J.; Albacete, A.; Faize, L.; Faize, M.; Pérez-Alfocea, F.; Hernández, J.A. Interaction between hydrogen peroxide and plant hormones during germination and the early growth of pea seedlings. *Plant Cell Environ.* **2010**, *33*, 981–994. [CrossRef]
62. Barba-Espin, G.; Nicolas, E.; Almansa, M.S.; Cantero-Navarro, E.; Albacete, A.; Hernández, J.A.; Díaz-Vivancos, P. Role of thioproline on seed germination: Interaction ROS-ABA and effects on antioxidative metabolism. *Plant Physiol. Biochem.* **2012**, *59*, 30–36. [CrossRef]
63. Kozaki, A.; Aoyanagi, T. Molecular Aspects of Seed Development Controlled by Gibberellins and Abscisic Acids. *Int. J. Mol. Sci.* **2022**, *23*, 1876. [CrossRef]
64. Hedden, P. The Current Status of Research on Gibberellin Biosynthesis. *Plant Cell Physiol.* **2020**, *61*, 1832–1849. [CrossRef]
65. Frey, A.; Effroy, D.; Lefebvre, V.; Seo, M.; Perreau, F.; Berger, A.; Sechet, J.; To, A.; North, H.M.; Marion-Poll, A. Epoxycarotenoid cleavage by NCED5 fine-tunes ABA accumulation and affects seed dormancy and drought tolerance with other NCED family members. *Plant J.* **2012**, *70*, 501–512. [CrossRef] [PubMed]
66. Bulley, S.M.; Cooney, J.M.; Laing, W. Elevating Ascorbate in Arabidopsis Stimulates the Production of Abscisic Acid, Phaseic Acid, and to a Lesser Extent Auxin (IAA) and Jasmonates, Resulting in Increased Expression of DHAR1 and Multiple Transcription Factors Associated with Abiotic Stress Tolerance. *Int. J. Mol. Sci.* **2021**, *22*, 6743. [PubMed]
67. Kobrehel, K.; Wong, J.H.; Balogh, A.; Kiss, F.; Yee, B.C.; Buchanan, B.B. Specific reduction of wheat storage proteins by thioredoxin h. *Plant Physiol.* **1992**, *99*, 919–924. [CrossRef] [PubMed]
68. Oliver, M.J.; Farrant, J.M.; Hilhorst, H.W.M.; Mundree, S.; Williams, B.; Bewley, J.D. Desiccation Tolerance: Avoiding Cellular Damage During Drying and Rehydration. *Annu. Rev. Plant. Biol.* **2020**, *71*, 435–460. [CrossRef]
69. Sergeant, M.J.; Li, J.J.; Fox, C.; Brookbank, N.; Rea, D.; Bugg, T.D.; Thompson, A.J. Selective inhibition of carotenoid cleavage dioxygenases: Phenotypic effects on shoot branching. *J. Biol. Chem.* **2009**, *284*, 5257–5264. [CrossRef] [PubMed]
70. Nakata, P.A. Influence of calcium oxalate crystal accumulation on the calcium content of seeds from Medicago truncatula. *Plant Sci.* **2012**, *185*, 246–249. [CrossRef] [PubMed]
71. Minor, E.A.; Court, B.L.; Young, J.I.; Wang, G. Ascorbate induces ten-eleven translocation (Tet) methylcytosine dioxygenase-mediated generation of 5-hydroxymethylcytosine. *J. Biol. Chem.* **2013**, *288*, 13669–13674. [CrossRef] [PubMed]
72. Kawakatsu, T.; Nery, J.R.; Castanon, R.; Ecker, J.R. Dynamic DNA methylation reconfiguration during seed development and germination. *Genome. Biol.* **2017**, *18*, 171. [CrossRef]
73. Chung, T.L.; Brena, R.M.; Kolle, G.; Grimmond, S.M.; Berman, B.P.; Laird, P.W.; Pera, M.F.; Wolvetang, E.J. Vitamin C promotes widespread yet specific DNA demethylation of the epigenome in human embryonic stem cells. *Stem. Cells* **2010**, *28*, 1848–1855. [CrossRef]
74. Jeevan Kumar, S.P.; Rajendra Prasad, S.; Banerjee, R.; Thammineni, C. Seed birth to death: Dual functions of reactive oxygen species in seed physiology. *Ann. Bot.* **2015**, *116*, 663–668. [CrossRef] [PubMed]
75. Farooq, M.A.; Zhang, X.; Zafar, M.M.; Ma, W.; Zhao, J. Roles of Reactive Oxygen Species and Mitochondria in Seed Germination. *Front. Plant Sci.* **2021**, *12*, 781734. [CrossRef] [PubMed]
76. Katsuya-Gaviria, K.; Caro, E.; Carrillo-Barral, N.; Iglesias-Fernández, R. Reactive Oxygen Species (ROS) and Nucleic Acid Modifications during Seed Dormancy. *Plants* **2020**, *9*, 679. [CrossRef] [PubMed]
77. Jurdak, R.; Rodrigues, G.A.G.; Chaumont, N.; Schivre, G.; Bourbousse, C.; Barneche, F.; Bou Dagher Kharrat, M.; Bailly, C. Intracellular reactive oxygen species trafficking participates in seed dormancy alleviation in Arabidopsis seeds. *New Phytol.* **2022**, *234*, 850–866. [CrossRef]
78. Iwasaki, M.; Penfield, S.; Lopez-Molina, L. Parental and Environmental Control of Seed Dormancy in Arabidopsis thaliana. *Annu. Rev. Plant Biol.* **2022**, *73*, 355–378. [CrossRef]
79. Carrillo-Barral, N.; Rodríguez-Gacio, M.d.C.; Matilla, A.J. Delay of Germination-1 (DOG1): A Key to Understanding Seed Dormancy. *Plants* **2020**, *9*, 480. [CrossRef]
80. Chen, F.; Li, Y.; Li, X.; Li, W.; Xu, J.; Cao, H.; Wang, Z.; Li, Y.; Soppe, W.J.J.; Liu, Y. Ectopic expression of the Arabidopsis florigen gene FLOWERING LOCUS T in seeds enhances seed dormancy via the GA and DOG1 pathways. *Plant J.* **2021**, *107*, 909–924. [CrossRef]

81. Chen, M.; Penfield, S. Feedback regulation of COOLAIR expression controls seed dormancy and flowering time. *Science* **2018**, *360*, 1014–1017. [CrossRef]
82. Müller, K.; Bouyer, D.; Schnittger, A.; Kermode, A.R. Evolutionarily conserved histone methylation dynamics during seed life-cycle transitions. *PLoS ONE* **2012**, *7*, e51532. [CrossRef]
83. Barth, C.; De Tullio, M.; Conklin, P.L. The role of ascorbic acid in the control of flowering time and the onset of senescence. *J. Exp. Bot.* **2006**, *57*, 1657–1665. [CrossRef]
84. Jurdak, R.; Launay-Avon, A.; Paysant-Le Roux, C.; Bailly, C. Retrograde signalling from the mitochondria to the nucleus translates the positive effect of ethylene on dormancy breaking of Arabidopsis thaliana seeds. *New Phytol.* **2021**, *229*, 2192–2205. [CrossRef] [PubMed]
85. Shanmugabalaji, V.; Chahtane, H.; Accossato, S.; Rahire, M.; Gouzerh, G.; Lopez-Molina, L.; Kessler, F. Chloroplast Biogenesis Controlled by DELLA-TOC159 Interaction in Early Plant Development. *Curr. Biol.* **2018**, *28*, 2616–2623.e2615. [CrossRef] [PubMed]
86. Luo, X.; Dai, Y.; Zheng, C.; Yang, Y.; Chen, W.; Wang, Q.; Chandrasekaran, U.; Du, J.; Liu, W.; Shu, K. The ABI4-RbohD/VTC2 regulatory module promotes reactive oxygen species (ROS) accumulation to decrease seed germination under salinity stress. *New Phytol.* **2021**, *229*, 950–962. [CrossRef] [PubMed]
87. Ye, N.; Zhu, G.; Liu, Y.; Zhang, A.; Li, Y.; Liu, R.; Shi, L.; Jia, L.; Zhang, J. Ascorbic acid and reactive oxygen species are involved in the inhibition of seed germination by abscisic acid in rice seeds. *J. Exp. Bot.* **2012**, *63*, 1809–1822. [CrossRef] [PubMed]
88. Stanley, L.; Paolo, B.; Lukasz, W.; Szymon Kubala, S.; Roberta, P.; Katzarina, L.; Muriel, Q.; Malgorzata, G. Seed Priming: New Comprehensive Approaches for an Old Empirical Technique. In *New Challenges in Seed Biology*; Susana, A., Alma, B., Eds.; IntechOpen: Rijeka, Croatian, 2016; Chapter 1. [CrossRef]
89. Paparella, S.; Araújo, S.S.; Rossi, G.; Wijayasinghe, M.; Carbonera, D.; Balestrazzi, A. Seed priming: State of the art and new perspectives. *Plant Cell. Rep.* **2015**, *34*, 1281–1293. [CrossRef]
90. Zulfiqar, F.; Nafees, M.; Chen, J.; Darras, A.; Ferrante, A.; Hancock, J.T.; Ashraf, M.; Zaid, A.; Latif, N.; Corpas, F.J.; et al. Chemical priming enhances plant tolerance to salt stress. *Front. Plant Sci.* **2022**, *13*, 946922. [CrossRef]
91. Ibrahim, E.A. Seed priming to alleviate salinity stress in germinating seeds. *J. Plant Physiol.* **2016**, *192*, 38–46. [CrossRef]
92. Farooq, M.; Tabassum, R.; Afzal, I. Enhancing the performance of direct seeded fine rice by seed priming. *Plant Prod. Sci.* **2006**, *9*, 446–456. [CrossRef]
93. Jafar, M.Z.; Farooq, M.; Cheema, M.A.; Afzal, I.; Basra, S.M.A.; Wahid, M.A.; Aziz, T.; Shahid, M. Improving the Performance of Wheat by Seed Priming Under Saline Conditions. *J. Agron. Crop Sci.* **2012**, *198*, 38–45. [CrossRef]
94. Shah, T.; Latif, S.; Khan, H.; Munsif, F.; Nie, L.X. Ascorbic Acid Priming Enhances Seed Germination and Seedling Growth of Winter Wheat under Low Temperature Due to Late Sowing in Pakistan. *Agronomy* **2019**, *9*, 757. [CrossRef]
95. Ahmad, I.; Khaliq, T.; Ahmad, A.; Basra, S.M.; Hasnain, Z.; Ali, A. Effect of seed priming with ascorbic acid, salicylic acid and hydrogen peroxide on emergence, vigor and antioxidant activities of maize. *Afr. J. Biotechnol.* **2012**, *11*, 1127–1132.
96. Farooq, M.; Irfan, M.; Aziz, T.; Ahmad, I.; Cheema, S.A. Seed Priming with Ascorbic Acid Improves Drought Resistance of Wheat. *J. Agron. Crop Sci.* **2013**, *199*, 12–22. [CrossRef]
97. Alcantara, B.K.; Rizzi, V.; Gaziola, S.A.; Azevedo, R.A. Soluble amino acid profile, mineral nutrient and carbohydrate content of maize kernels harvested from plants submitted to ascorbic acid seed priming. *An. Acad. Bras. Cienc.* **2017**, *89*, 695–704. [CrossRef] [PubMed]
98. Baig, Z.; Khan, N.; Sahar, S.; Sattar, S.; Zehra, R. Effects of seed priming with ascorbic acid to mitigate salinity stress on three wheat (*Triticum aestivum* L.) cultivars. *Acta Ecologica Sinica* **2021**, *41*, 491–498. [CrossRef]
99. Alves, R.D.; Rossatto, D.R.; da Silva, J.D.; Checchio, M.V.; de Oliveira, K.R.; Oliveira, F.D.; de Queiroz, S.F.; da Cruz, M.C.P.; Gratao, P.L. Seed priming with ascorbic acid enhances salt tolerance in micro-tom tomato plants by modifying the antioxidant defense system components. *Biocatal. Agric. Biotechnol.* **2021**, *31*, 101927. [CrossRef]
100. Kasim, W.A.; Nessem, A.A.; Gaber, A. Alleviation of Drought Stress in *Vicia faba* by Seed Priming with Ascorbic Acid or Extracts of Garlic and Carrot. *Egypt. J. Bot.* **2017**, *57*, 45–59. [CrossRef]
101. Kasim, W.A.; Nessem, A.A.; Gaber, A. Effect of seed priming with aqueous extracts of carrot roots, garlic cloves or ascorbic acid on the yield of *Vicia faba* grown under drought stress. *Pak. J. Bot.* **2019**, *51*, 1979–1985. [CrossRef]
102. Shahnawaz, M.; Sanadhya, D. Aluminium induced oxidative stress and antioxidants system in two barley varieties and its alleviation through ascorbic acid and salicylic acid seed priming approach. *Int. J. Life Sci. Pharma Res.* **2017**, *7*, L26–L37.
103. Xia, F.; Cheng, H.; Chen, L.; Zhu, H.; Mao, P.; Wang, M. Influence of exogenous ascorbic acid and glutathione priming on mitochondrial structural and functional systems to alleviate aging damage in oat seeds. *BMC Plant Biol.* **2020**, *20*, 104. [CrossRef]
104. Fercha, A.; Capriotti, A.L.; Caruso, G.; Cavaliere, C.; Samperi, R.; Stampachiacchiere, S.; Laganà, A. Comparative analysis of metabolic proteome variation in ascorbate-primed and unprimed wheat seeds during germination under salt stress. *J. Proteom.* **2014**, *108*, 238–257. [CrossRef]
105. Munns, R.; Gilliham, M. Salinity tolerance of crops—What is the cost? *New Phytol.* **2015**, *208*, 668–673. [CrossRef] [PubMed]
106. Gulyas, L.; Powell, J.R. Predicting the Future: Parental Progeny Investment in Response to Environmental Stress Cues. *Front. Cell Dev. Biol.* **2019**, *7*, 115. [CrossRef] [PubMed]

107. Segerstrom, S.C. Resources, stress, and immunity: An ecological perspective on human psychoneuroimmunology. *Ann. Behav. Med.* **2010**, *40*, 114–125. [CrossRef] [PubMed]
108. Richards, C.L.; Alonso, C.; Becker, C.; Bossdorf, O.; Bucher, E.; Colomé-Tatché, M.; Durka, W.; Engelhardt, J.; Gaspar, B.; Gogol-Döring, A.; et al. Ecological plant epigenetics: Evidence from model and non-model species, and the way forward. *Ecol. Lett.* **2017**, *20*, 1576–1590. [CrossRef] [PubMed]

MDPI AG
Grosspeteranlage 5
4052 Basel
Switzerland
Tel.: +41 61 683 77 34

Seeds Editorial Office
E-mail: seeds@mdpi.com
www.mdpi.com/journal/seeds

www.ingramcontent.com/pod-product-compliance
Lightning Source LLC
Chambersburg PA
CBHW041342210926
51850CB00013B/2953
* 9 7 8 3 7 2 5 8 2 8 3 5 7 *